INSTITUTE OF GEOLOGICAL SCIENCES

Natural Environment Research Council

Mineral Assessment Report 33

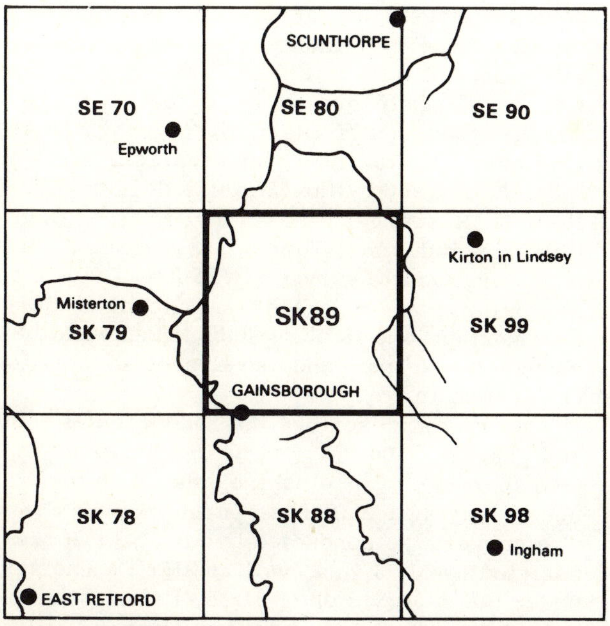

The sand and gravel resources of the country north of Gainsborough, Lincolnshire

Description of 1:25 000 resource sheet SK 89

J. R. Gozzard and D. Price

ISBN 0 11 884033 9

London Her Majesty's Stationery Office 1978

The first twelve reports on the assessment of British sand and gravel resources appeared in the Report Series of the Institute of Geological Sciences as a subseries. Report No. 13 onwards are appearing in the Mineral Assessment Report Series of the Institute. Details of published reports appear at the end of this Report.

Any enquiries concerning this report may be addressed to Head, Industrial Minerals Assessment Unit, Institute of Geological Sciences, Keyworth, Nottingham NG12 5GG

PREFACE

National resources of many industrial minerals may seem so large that stocktaking appears unnecessary, but the demand for minerals and for land for all purposes is intensifying and it has become increasingly clear in recent years that regional assessments of the resources of these minerals should be undertaken. The publication of information about the quantity and quality of deposits over large areas is intended to provide a comprehensive factual background against which planning decisions can be made.

Sand and gravel, considered together as naturally occurring aggregate, was selected as the bulk mineral demanding the most urgent attention, initially in the south-east of England, where about half the national output is won and very few sources of alternative aggregates are available. Following a short feasibility project, initiated in 1966 by the Ministry of Land and National Resources, the Mineral Assessment Unit (now Industrial Minerals Assessment Unit) began systematic surveys in 1968. The work is now being financed by the Department of the Environment and is being undertaken with the cooperation of the Sand and Gravel Association of Great Britain.

This report describes the resources of sand and gravel of 100 km^2 of country north of Gainsborough, Lincolnshire, shown on the accompanying 1:25 000 resource map SK 89. The survey was conducted by Mr J. R. Gozzard, assisted by Mr J.H. Lovell and Mr D. Thomas, under the supervision of Mr D. Price. The report has been written by Mr. Gozzard and Mr. Price. The work, which was controlled from the sub-unit based in Leeds (J. H. Hull, Officer-in-Charge), is based on six-inch scale geological surveys carried out by Mr G. H. Rhys and Mr E. G. Smith in 1960-65 (published in part on New Series one-inch sheets 88 (Doncaster) and 101 (East Retford)) and by Mr R. J. Bull in 1972-73 (to be published on New Series Sheet 89 (Brigg)), together with information from a one-inch scale survey published in 1886 on Old Series Sheet 83.

Mr. J. W. Gardner, CBE (Land Agent) negotiated access to land for drilling. The ready cooperation of land owners and tenants is gratefully acknowledged.

Austin W. Woodland
Director

Institute of Geological Sciences
Exhibition Road
South Kensington
London SW7 2DE

19 September, 1978

CONTENTS

FIGURES

TABLES

The sand and gravel resources of the country north of Gainsborough, Lincolnshire

Description of 1:25 000 resource sheet SK 89

J. R. Gozzard and D. Price

SUMMARY

The geological maps of the Institute of Geological Sciences, pre-existing borehole information, and 62 boreholes drilled for the Industrial Minerals Assessment Unit, form the basis of the assessment of sand and gravel resources in the area north of Gainsborough, Lincolnshire.

All deposits in the area which might be potentially workable for sand and gravel have been investigated and a simple statistical method has been used to estimate the volume. The reliability of the volume estimates is given at the symmetrical 95 per cent probability level.

The 1:25 000 map is divided into five resource blocks, containing between 7.7 and 15.9 km^2 of sand and gravel. For each block the geology of the deposits is described and the mineral-bearing area, the mean thicknesses of overburden and mineral and the mean gradings are stated. Detailed borehole data are also given. The geology, the position of the boreholes and the outlines of the resource blocks are shown on the accompanying map.

Bibliographical reference

GOZZARD, J. R. and PRICE, D. 1978. The sand and gravel resources of the country north of Gainsborough, Lincolnshire: Description of 1:25 000 resource sheet SK 89. Miner. Assess. Rep. Inst. Geol. Sci., No. 33.

Authors

J. R. Gozzard, B. Sc. and D. Price, B. Sc.
Institute of Geological Sciences, Keyworth, Nottingham NG12 5GG

INTRODUCTION

The survey is concerned with the estimation of resources, which include deposits that are not currently exploitable but have a foreseeable use, rather than reserves, which can only be assessed in the light of current, locally prevailing economic considerations. Clearly, both the economic and the social factors used to decide whether a deposit may be workable in the future cannot be predicted; they are likely to change with time. Deposits not currently economically workable may be exploited as demand increases, as higher grade or alternative materials become scarce, or as improved processing techniques are applied to them. The improved knowledge of the main physical properties of the resource and their variability, which this survey seeks to provide, will add significantly to the factual background against which planning policies can be decided (Archer, 1969; Thurrell, 1971; Harris and others, 1974).

The survey provides information at the 'indicated' level "for which tonnage and grade are computed partly from specific measurements, samples or production data and partly from projection for a reasonable distance on geological evidence. The sites available for inspection, measurement, and sampling are too widely spaced to permit the mineral bodies to be outlined completely or the grade established throughout" (Bureau of Mines and Geological Survey, 1948, p. 15).

It follows that the whereabouts of reserves must still be established and their size and quality proved by the customary detailed exploration and evaluation undertaken by the industry. However, the information provided by this survey should assist in the selection of the best targets for such further work.

The following arbitrary physical criteria have been adopted:

a. The deposit should average at least 1 m in thickness.
b. The ratio of overburden to sand and gravel should be no more than 3:1.
c. The proportion of fines (particles passing the No. 240 mesh BS sieve, about 1/16 mm) should not exceed 40 per cent.
d. The deposit must lie within 25 m of the surface, this being taken as the likely maximum working depth under most circumstances.
It follows from the second criterion that boreholes are drilled no deeper than 18 m if no sand and gravel has been proved.

A deposit of sand and gravel which broadly meets these criteria is regarded as 'potentially workable'

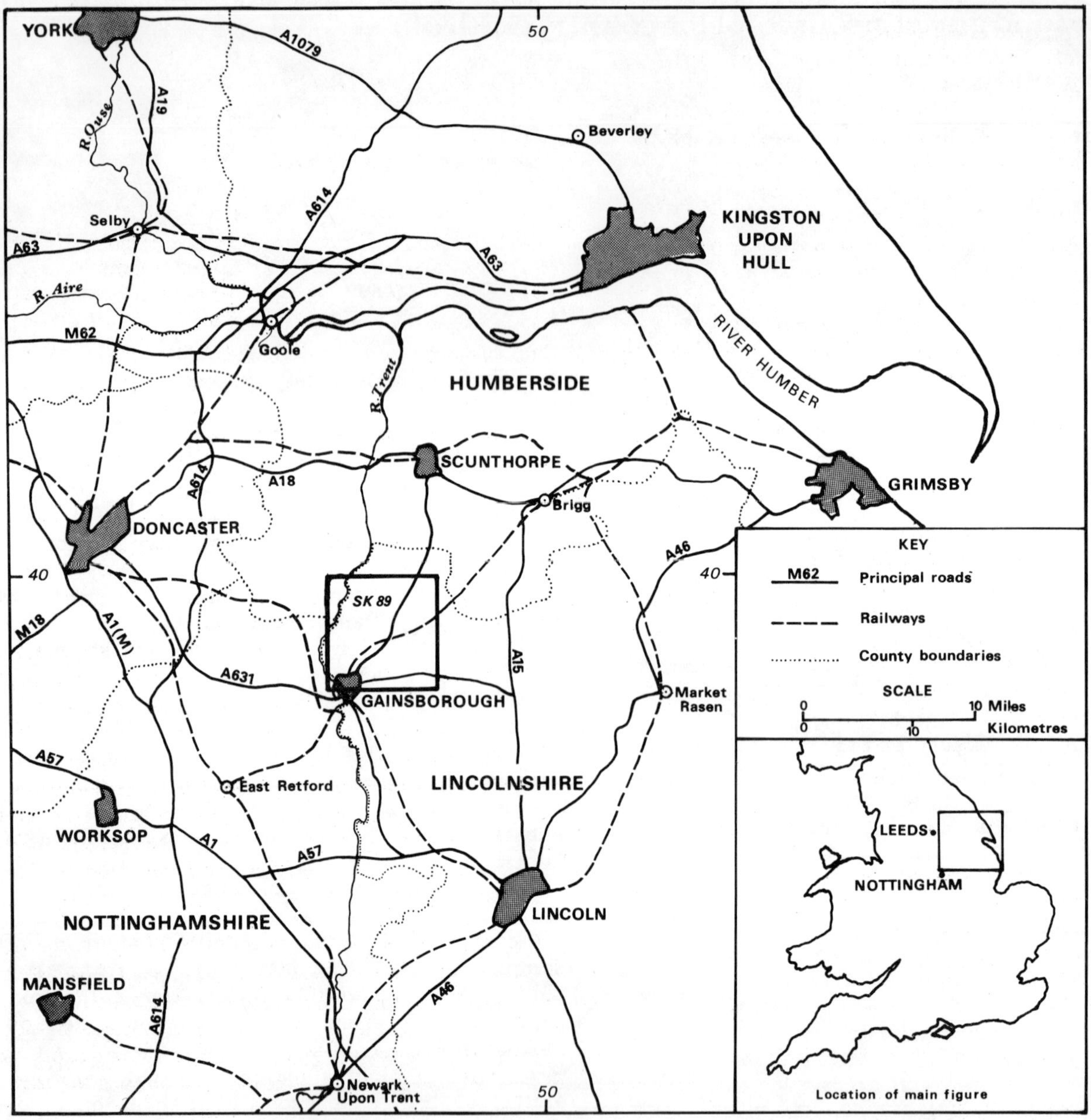

Figure 1. Map showing the location of sheet SK 89

and is described and assessed as 'mineral' in this report. As the assessment is at the indicated level, parts of such a deposit may not satisfy all the criteria.

For the particular needs of assessing sand and gravel resources, a grain-size classification based on the geometric scale 1/16 mm, ¼ mm, 1 mm, 4 mm, 16 mm has been adopted. The boundaries between fines (that is, the clay and silt fractions) and sand, and between sand and gravel-grade material, are placed at 1/16 mm and 4 mm respectively (see Appendix C).

The volume and other characteristics are assessed within resource blocks, each of which, ideally, contains approximately 10 km^2 of sand and gravel. No account is taken of any factors, for example, roads, villages and high agricultural or landscape value, which might stand in the way of sand and gravel being exploited, although towns are excluded. The estimated total volume therefore bears no simple relationship to the amount that could be extracted in practice.

> It must be emphasised that the assessment applies to the resource block as a whole. Valid conclusions cannot be drawn about the mineral in parts of a block, except in the immediate vicinity of the actual sample points.

DESCRIPTION OF RESOURCE SHEET

GENERAL

The district described (Fig. 1) lies to the north and north-east of Gainsborough and is largely within the valley of the Trent. The broad floodplain of the river is bordered by the First Terrace of the Trent, up to about 3 km wide; to the east of the river gently undulating ground rises to a little over 38 m above OD immediately north-east of Gainsborough and at Hardwick Hill [840 995] in the north. The eastern part of the higher ground is drained by the River Eau which joins the Trent some distance to the north.

Gainsborough is a commercial, residential and industrial centre but the remainder of the area is devoted to agriculture and forestry.

GEOLOGY

The geological sequence is summarised in Table 1. The deposits are listed as far as possible in order of increasing age and their relationship is illustrated in the schematic cross-sections (Fig. 2).

SOLID

Keuper Marl

The upper part of the Keuper Marl crops out in the western part of the district but elsewhere it is almost entirely concealed by superficial deposits. It consists mainly of reddish brown mudstones and siltstones with thin beds of fine-grained dolomitic sandstone, but the uppermost beds, the Tea Green Marl, are greenish grey mudstones.

Table 1. Stratigraphy

DRIFT	
Recent and Pleistocene	Peat
	Alluvium
	Blown Sand
	First Terrace of the River Trent
	Terrace of the River Eau
	25-Foot Drift of the Vale of York
	Older River Sand and Gravel
	Head
	Glacial Sand and Gravel
	Boulder Clay
SOLID	
Jurassic	Lias
Triassic	Rhaetic (Penarth Group)
	Keuper Marl (Mercia Mudstone Group)

Rhaetic

The Rhaetic strata, about 18 m thick, consist of black fissile mudstones (Westbury Beds) overlain by grey and reddish brown mudstones with sporadic beds of argillaceous limestone (Cotham Beds).

Lias

Dark grey shales and mudstones of Lower Jurassic age succeed the Rhaetic strata and crop out in the eastern part of the district. They include thin beds of argillaceous limestone, and fossils, including bivalves and crinoids, are abundant.

DRIFT

Boulder Clay

Boulder Clay covers most of the eastern part of the district and is also found immediately east of Gainsborough. It consists of brown clay and silt with erratic pebbles and cobbles of flint, chalk, quartz, quartzite, siltstone and sandstone.

Glacial Sand and Gravel

Glacial Sand and Gravel, occurring between Hardwick Hill [840 995] and Laughton [848 973] and around Blyton, consist of 'clayey' sands (see Appendix C) and gravels up to 4.4 m thick; they generally rest on bedrock but are in places underlain by boulder clay. Sand and 'clayey' sand up to 5 m thick are found within Boulder Clay at Corringham.

Head

Soliflucted red, brown and grey clays with brown and grey sands floor a narrow valley at the foot of the Rhaetic escarpment east of Gainsborough.

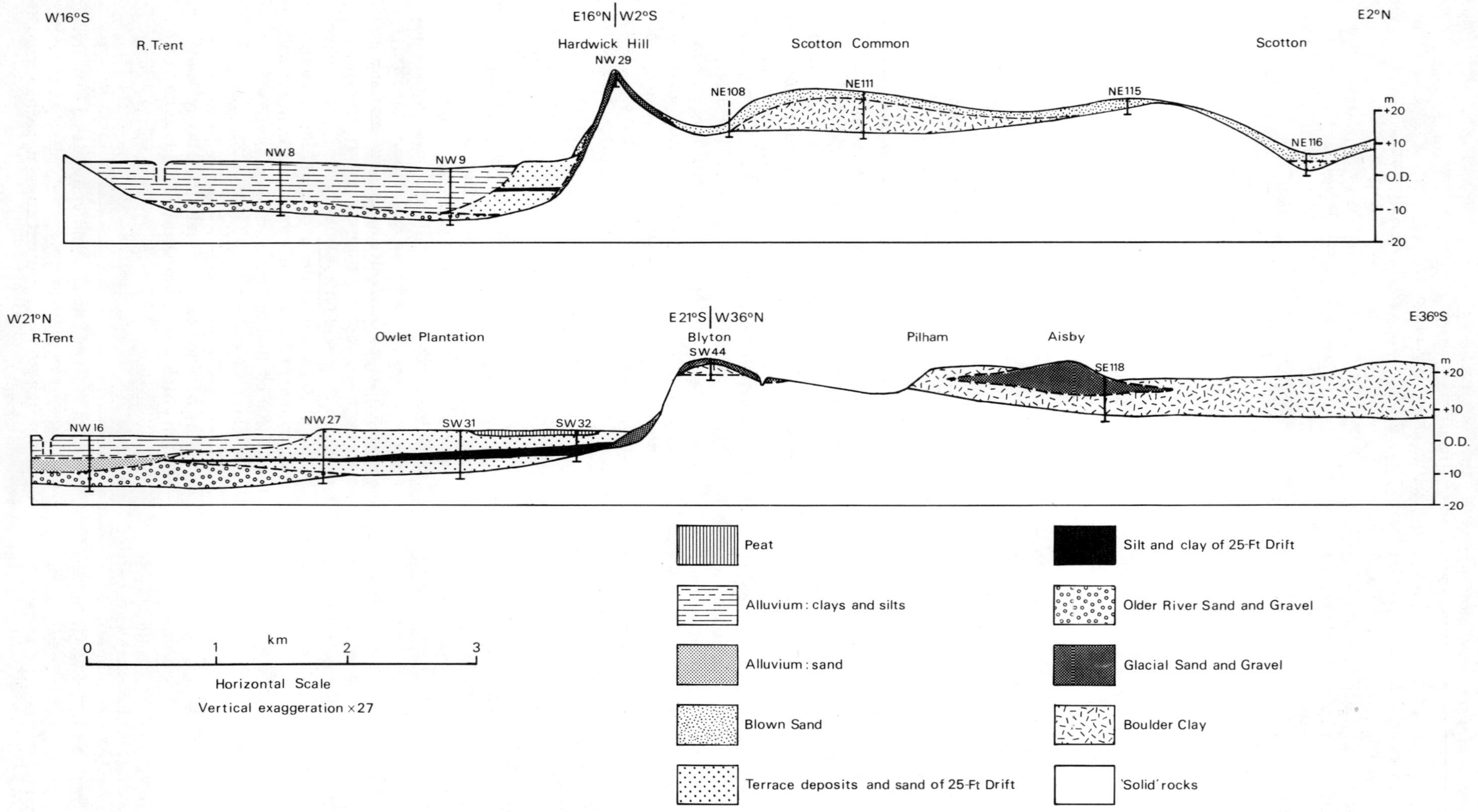

Figure 2. Horizontal sections illustrating the Drift geology of the district (the lines of section are marked on the resource sheet)

Older River Sand and Gravel

Sand and gravel which appear to occupy a buried channel following approximately the present course of the river have been proved beneath younger deposits by boreholes in the Trent Valley. They are up to 7.5 m thick. There is no evidence of their age.

25-Foot Drift of the Vale of York

Beneath the First Terrace deposits of the Trent north of Gainsborough, sands with little or no gravel rest on bedrock or Older River Sand and Gravel. They are thought to be related to the marginal and lower sands of the 25-Foot Drift of the Vale of York[1]. They are commonly overlain by thin silts and clays which are correlated with the silt and clay of the 25-Ft Drift. These sands, silts and clays, none of which reach the surface in the Trent Valley, are considered to be associated with the low-level phase of Devensian Lake Humber (Gaunt and others, 1971, p. 281; Gaunt, 1974, p. 195).

First Terrace of the River Trent

This terrace is present at the surface north of Gainsborough in a strip up to 3 km wide bordering the floodplain. The deposits forming it continue extensively beneath alluvium and locally protrude through the floodplain as sandy hillocks. They consist of at least 9.4 m of sand with a few small pebbles, and overlie the 25-Ft Drift. The sand is thought to have been deposited by the River Trent following the drainage of Lake Humber and is thus comparable in origin to some of the sand of the 25-Ft Drift (as suggested by Smith and others, 1973, p. 219).

Terrace of the River Eau

This terrace is of limited extent, being found only at Northorpe and to the east of Scotton. The deposits forming it were encountered beneath alluvium in a borehole near the latter place and consist of sandy gravel with flint and quartzite pebbles.

Blown Sand

Blown Sand is extensive in the northern part of the district, where it is banked against the Rhaetic escarpment and blankets Scotton Common; small patches are also found west and north-west of Wildsworth. It generally consists of 'clayey' well rounded sand and has a maximum proved thickness of 4.6 m. The deposit is the result of the redistribution of terrace sand and sand of the 25-Ft Drift from late Devensian times onwards.

Alluvium

In late Devensian times a rapid fall in sea level led to incision by the Trent (Gaunt and others, 1971, p. 282). Alluvial deposition in the incised channel began in the Flandrian. Sands with some pebbles were deposited in the channel and, in the form of bank deposits, against its eastern wall. There followed the deposition of clays and silts with some interbedded peats, which attained thicknesses of up to 12.2 m and continued to form until the construction of artificial flood banks. In places warping, artificially induced sedimentation, has added to the accumulation of clay and silt.

Patches of alluvium are common on the First Terrace of the Trent; some of the more extensive of these are probably the result of warping.

Peat

Peat, up to at least 1.5 m thick, and in places containing lumps of calcareous tufa, covers a large area of the First Terrace between Blyton and Gainsborough, and occurs locally at the surface on the floodplain. It is also interbedded with the alluvial clays and silts.

COMPOSITION OF THE SAND AND GRAVEL

Glacial Sand and Gravel

These deposits have a mean grading of 10 per cent fines, 85 per cent sand and 5 per cent gravel but exhibit marked lateral variation (see description of block E). Subrounded quartz and quartzite dominate the gravel fraction, with angular to subrounded flint, chert and siltstone present in lesser amounts. The sand fraction is usually medium grained and comprises quartz with subordinate amounts of quartzite, flint, chert and other rock fragments.

Older River Sand and Gravel

These deposits, again, exhibit considerable lateral variation. Beneath the Trent floodplain, gravel generally accounts for between 27 and 41 per cent of the deposit and the mean grading is 2 per cent fines, 68 per cent sand and 30 per cent gravel. Elsewhere gravel is less common and the deposit has a mean grading of 2 per cent fines, 85 per cent sand and 13 per cent gravel. The gravel fraction consists predominantly of rounded quartz and quartzite, with lesser amounts of subangular flint and chert, subordinate amounts of sandstone, mudstone and limestone and rare ironstone and coal. The sand fraction is commonly medium grained and comprises quartz with some quartzite, chert and other rock fragments.

Sands of the 25-Ft Drift

Deposits here equated with the basal and marginal sands of the 25-Ft Drift consist of fine to medium sands with only a little gravel; they have a mean grading of 4 per cent fines, 96 per cent sand and a trace of gravel. Their composition is similar to that of the Older River Sand and Gravel.

First Terrace of the River Trent

This deposit consists of sands with sporadic thin seams of pebbles and has a mean grading of 5 per cent fines, 95 per cent sand and a trace of gravel. Pebbles comprise rounded quartz and quartzite with sporadic subangular chert. The sand fraction is similar to that of the Older River Sand and Gravel but it may include a higher proportion of coal.

1 'Hereinafter abbreviated to '25-Ft Drift'

Terrace of the River Eau

This terrace, proved by only one borehole (NE 116), consists of sandy gravel with a mean grading of 2 per cent fines, 54 per cent sand and 44 per cent gravel. The gravel fraction consists of subrounded quartz and quartzite with subangular chert and flint, some rounded sandstone and many fossil shells. The sand fraction is poorly sorted and comprises quartz with quartzite and other lithic fragments.

Blown Sand

As its name suggests this deposit is entirely pebble-free, having a mean grading of fines 12 per cent and sand 88 per cent. The sand comprises fine- to medium- grained, subrounded to well rounded quartz with lesser amounts of quartzite, chert and other lithic grains.

Alluvium

Alluvial mineral consists of sands with sporadic pebbly layers and is distinguished from the older First Terrace deposits by its greater pebble content and smaller fines content. Its mean grading is fines 4 per cent, sand 92 per cent and gravel 4 per cent. The composition of the sand and gravel fractions is very similar to that of the Older River Sand and Gravel and of the First Terrace of the River Trent.

THE MAP

The sand and gravel resource map is folded into the pocket at the end of this report. The base map is the Ordnance Survey 1:25 000 Outline Edition in grey, on which the topography is shown by contours in green, the geological data in black and the mineral resource information in red.

Geological data

The geological boundary lines are based on six-inch geological surveys made between 1960 and 1965 and published on the one-inch scale on sheets 88 (Doncaster) and 101 (East Retford), together with surveys in 1972-73 (as yet unpublished) and a one-inch geological survey published in 1886. Borehole data, which include the stratigraphical relations and mean particle-size distributions of sand and gravel samples collected during the assessment survey, are also shown.

The geological boundaries are the best available interpretation of information available at the time of the survey. However, it is inevitable that local irregularities or discrepancies will be revealed by some boreholes; these are taken into account in the assessment of resources.

Mineral resource information

The mineral-bearing ground is sub-divided into resource blocks (see Appendix A). Within a resource block the mineral is sub-divided into areas where it is 'exposed' and areas where it is present in continuous (or almost continuous) spreads beneath overburden. The mineral is identified as 'exposed' where overburden, commonly consisting only of soil and subsoil, averages less than 1.0 m in thickness.

Areas where bedrock outcrops or where evidence indicates the absence of potentially workable sand and gravel are uncoloured on the map. In such areas it has been assumed that the mineral is absent except, possibly, in infrequent and relatively minor patches, which can neither be outlined nor assessed in the context of this survey. Areas of unassessed sand and gravel, for example in built-up areas, are indicated by a red stipple.

For the most part the depicted distribution of the various categories of deposits is based on the mapped geological boundaries. Where there is transition from one category to another, which cannot be related to the geological map and which cannot be delineated accurately, inferred boundaries, shown by a distinctive symbol, have been inserted. The symbol is intended to convey an approximate location within a likely zone of occurrence rather than represent the breadth of the zone, its size being limited only by cartographic considerations. For the purpose of measuring area the centre line of the symbol is used.

RESULTS

The statistical results of the survey are summarised in Table 2. Further grading particulars are given in Figs. 3 to 8 and Tables 3 to 7. All limits quoted in this report have been calculated at the symmetrical 95 per cent probability level.

Accuracy of results

For the five resource blocks the accuracy of the results at the 95 per cent probability level varies between 8 per cent and 41 per cent (that is, it is probable that nineteen times out of twenty the true volumes present lie within these limits). However, the true values are more likely to be nearer the figures estimated than the limits. Moreover, it is probable that in each block approximately the same percentage limits would apply for the estimate of volume of a very much smaller parcel of ground (say, 1 km^2) containing similar sand and gravel deposits if the results from the same number of sample points (as provided by, say, ten boreholes) were used in the calculation. Thus, if closer limits are needed for the quotation of reserves of part of a block it can be expected that data from more than ten sample points will be required, even if the area is quite small. This point can be illustrated by considering the whole of the potentially workable sand and gravel on this sheet. The volume (321 million m^3) can be estimated to limits of $\pm$ 19 per cent at the 95 per cent probability level, by a calculation based on the data from 59 sample points spread across the five resource blocks. However, it must be emphasised that the quoted volume of sand and gravel has no simple relationship with the amount that could be extracted in practice, as no allowance has been made in the calculations for any restraints (such as existing buildings and roads) on the use of the land for mineral working.

Table 2. The sand and gravel resources of Sheet SK 89

Resource Block	Area a (km^2)		Mean thickness (m)		Volume of mineral			Mean grading percentage		
						Limits at the 95% probability level		Fines	Sand	Gravel
	Block	Mineral	Overburden	Mineral	million m^3	± %	million m^3	-1/16mm	-4+1/16 mm	+4mm
A	16.9	15.9	6.0	7.5	119	30	36	3	80	17
B	7.7	7.7	0.5	12.7	98	8	8	3	94	3
C	11.1	10.2	0.6	5.5	56	41	23	5	93	2
B+C	18.8	17.9	0.6	8.6	154	24	37	4	94	2
D	10.1	9.5	0.3	2.7	26	37	10	11	86	3
E	51.5	9.4	1.2	2.6	24	39	9	10	85	5
A to E	97.3	52.7	2.3	6.1	321	19	6			

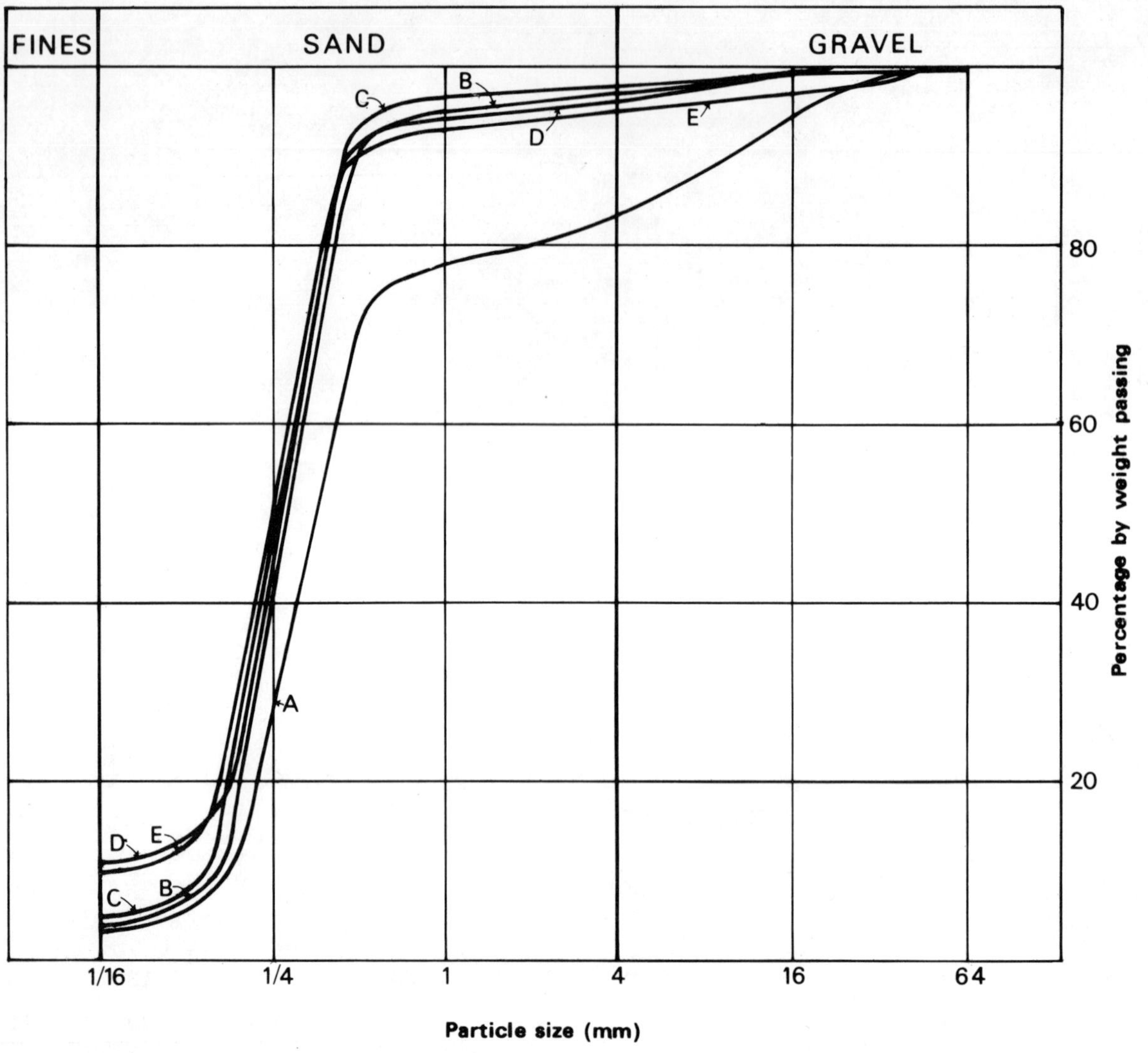

Percentages by weight

Resource Block	-1/16mm	+1/16mm -¼mm	+¼mm -1mm	+1mm -4mm	+4mm -16mm	+16mm
A	3	26	49	5	12	5
B	3	38	54	2	2	1
C	5	42	50	1	1	1
D	11	33	51	2	2	1
E	10	41	42	2	2	3

Figure 3. Mean particle-size distribution for the assessed thickness of sand and gravel in resource blocks A to E

Table 3. Block A: data from assessment boreholes

Borehole No.	Recorded thickness (m) Mineral	Overburden	Waste Partings[1]	Mean grading percentage Fines -1/16mm	Fine Sand $-\frac{1}{4}+1/16$mm	Medium Sand $-1+\frac{1}{4}$mm	Coarse Sand -4+1mm	Fine Gravel -16+4mm	Coarse Gravel +16mm
NW 4	Absent								
5	11.7	6.0		6	29	46	5	11	3
6	10.9	nil	1.1	5	43	50	1	1	
7	1.0	1.3		18	46	33	2	1	
8	3.8	11.3		3	10	73	5	7	2
9	(3.4)[2]	(12.2)		6	33	33	5	18	5
10	(1.0)	(8.0)		16	31	45	2	4	2
11	7.5	10.0		1	9	46	9	26	9
12	4.8	12.2		2	9	36	8	31	14
13	11.6	1.4		2	42	49	2	4	1
14	4.1	11.7		1	10	47	15	19	8
15	4.6	10.4		1	13	49	7	23	7
16	7.9	7.0		1	9	54	7	17	12
17	14.0	0.4	0.6	4	38	49	3	5	1
18	6.4	6.4		1	10	53	9	20	7
19	15.9	0.6	0.5	2	37	50	3	5	3
SW 27	9.9+	1.3	0.3	2	25	59	3	7	4
29	10.0	1.0		6	20	50	6	13	5
30	6.7	7.0	2.0	1	12	35	11	30	11

1. Seams of clay and/or silt overlain and underlain by mineral.

2. Brackets indicate that sand and gravel proved in the borehole does not meet criteria (a) and/or (b) of the definition of mineral (see p.1).

Table 4. Block B: data from assessment boreholes

Borehole No.	Recorded thickness (m) Mineral	Overburden	Waste Partings	Mean grading percentage Fines -1/16mm	Fine Sand $-\frac{1}{4}+1/16$mm	Medium Sand $-1+\frac{1}{4}$mm	Coarse Sand -4+1mm	Fine Gravel -16+4mm	Coarse Gravel +16mm
NW 20	12.9	nil	0.5	2	46	50	1	1	trace
22	13.0	0.3	0.7	3	42	53	1	1	
24	14.9	0.3		1	32	62	2	2	1
27	13.2	0.4	0.6	3	35	58	2	1	1
SW 28	12.7	1.8		4	38	54	2	2	
33	13.8	0.4	0.8	5	28	55	6	5	1
34	11.0	0.5	1.5	4	41	52	2	1	trace
36	12.6	0.4	1.8	3	39	50	3	4	1
37	10.6	0.4	0.7	6	48	45	1	trace	trace

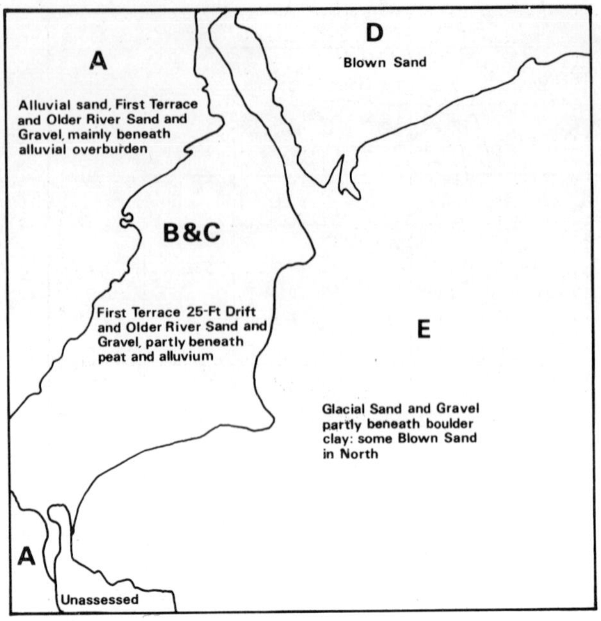

Figure 4. Drift geology summarised in relation to the resource block boundaries

NOTES ON RESOURCE BLOCKS

Block A is defined mainly by the extent of the Trent floodplain and is in two parts, the main one being north of Morton and the other immediately west of Gainsborough. Blocks B and C encompass the area occupied by the First Terrace of the River Trent north of Gainsborough; block B includes thick mineral deposits of relatively low variability, whereas in block C mineral is, in general, thinner and more variable. Block D includes the extensive blown sands around and west of Scotton. The remainder of the assessed area, containing deposits of Glacial Sand and Gravel, comprises block E.

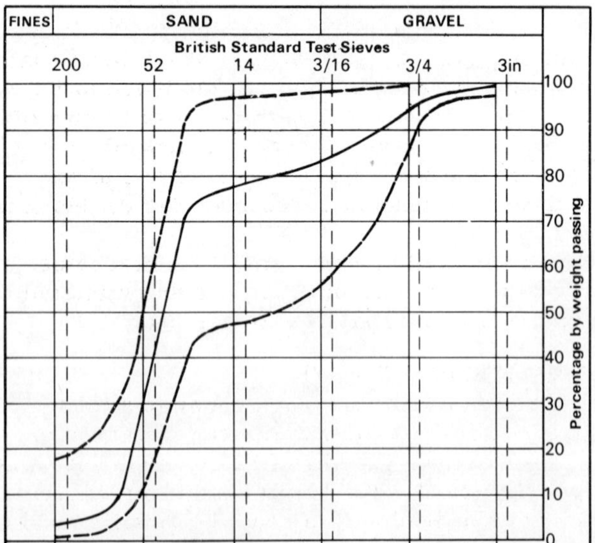

Figure 5. Grading characteristics of the mineral in block A: the continuous line represents the weighted mean grading of the block; the broken lines denote the envelope within which the mean grading curves for individual boreholes fall.

Block A (Table 3; Fig. 5)
In the extreme north-west of the block IMAU borehole NW 4 found no sand and gravel, and other borehole records and field observations confirm the absence of potentially workable sand and gravel hereabouts. An inferred boundary has, therefore, been inserted to delineate approximately the extent of barren ground. At boreholes NW 9 and NW 10 thickness of overburden is such as to render underlying pebbly sands not potentially workable but the area of 'barren' ground cannot be delineated on the resource map.

In the east of the main part of the block, First Terrace deposits rest on 25-Ft Drift clays and silts and Older River Sand and Gravel. Elsewhere in the block, all or part of this sequence was removed by Flandrian erosion and has been replaced by alluvial deposits. The distinction between the various sands and gravels is not everywhere clear, and their classification is tentative; deposits considered in this report to be alluvial may include terrace deposits and vice versa, and some of those regarded as Older River Sand and Gravel may post-date the Flandrian incision.

The mineral consists of sand, which is pebbly in places, in the Alluvium, terrace deposits and the 25-Ft Drift, and the more gravelly Older River Sand and Gravel. The last is absent north-west of the River Trent but elsewhere it ranges in thickness up to at least 9.0 m with a mean grading of 2 per cent fines, 68 per cent sand and 30 per cent gravel. Although the gravel content in individual boreholes ranges from 9 to 41 per cent, it generally lies between 27 and 35 per cent.

The younger sands are found mainly in the east and south-east of the main part of the block; they range in thickness up to at least 10.0 m and have a mean grading of 5 per cent fines, 94 per cent sand and 1 per cent gravel. Total thicknesses of mineral proved (Table 3) range from 1 to 15.9 m, and bearing in mind that borehole SW 27 did not bottom mineral, the mean thickness is 7.5 m+. The estimated volume of mineral in the block is 119 million $m^3 \pm 30$ per cent and its mean grading is 3 per cent fines, 80 per cent sand and 17 per cent gravel.

Four boreholes proved silt and clay waste partings ranging in thickness from 0.3 to 1.1 m. In a fifth borehole (SW 30) only dark peaty water was recovered between 8 and 10 m from surface and a 2-m waste parting of uncertain nature has been assumed to be present. The mean thickness of waste for the block is 0.3 m. Overburden, consisting of soil, silt and clay with some peat, varies considerably in thickness, with a mean of 6.0 m.

Block B (Table 4, Fig. 6)
This block encompasses the western part of the First Terrace of the River Trent north of Gainsborough. Mineral consists of First Terrace deposits overlying 25-Ft Drift sand and Older River Sand and Gravel and generally separated from them by thin silts and clays.

Older River Sand and Gravel was not found in boreholes NW 20 and SW 34 and 37 but was encountered in the other IMAU boreholes at depths

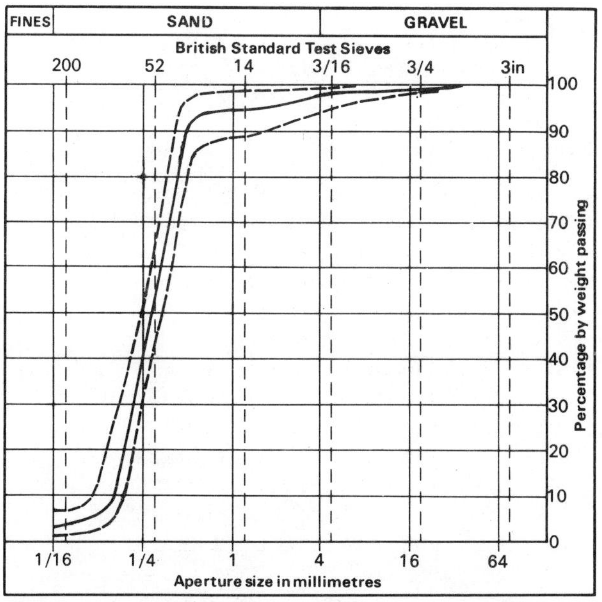

Figure 6. Grading characteristics of the mineral in block B (for explanation see Fig. 5)

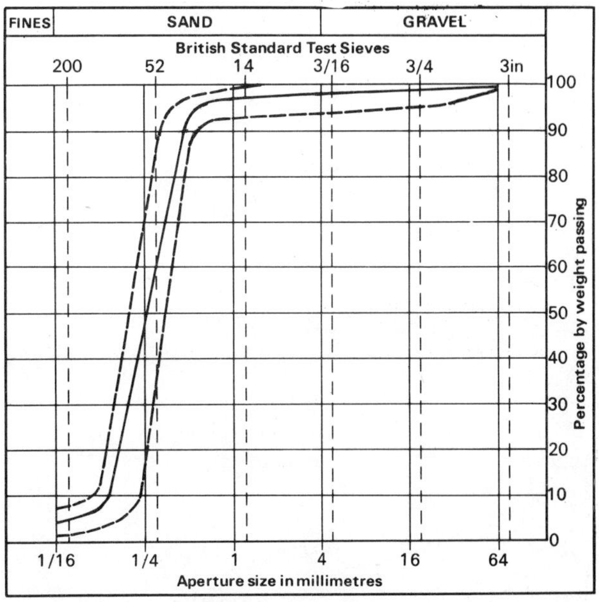

Figure 7. Grading characteristics of the mineral in block C (for explanation see Fig. 5)

between 10.8 and 13.0 m from the surface: it consists of from 1.0 to 4.0 m of pebbly sand with a mean grading of 1 per cent fines, 86 per cent sand and 13 per cent gravel. The overlying sand of the 25-Ft Drift ranges from 2.0 to 6.5 m in thickness and has a mean grading of 4 per cent fines, 95 per cent sand and 1 per cent gravel. The First Terrace deposits, from 4.5 to 9.0 m thick, consist of sands with a mean fines content of 4 per cent and only a trace of gravel.

Total mineral thicknesses in the block range from 10.6 to 14.9 m giving a mean of 12.7 m and an estimated volume of 98 million $m^3 \pm 8$ per cent. The overall mean grading is 3 per cent fines, 94 per cent sand and 3 per cent gravel.

The First Terrace is generally underlain by a clay or silt parting; it was not found at boreholes NW 24 and SW 28 but elsewhere proved thicknesses range from 0.5 to 1.8 m. In borehole SW 34 a 1.0-m clay parting was encountered in sand of the 25-Ft Drift. The mean thickness of waste in the block is 0.7 m. Overburden in the block usually consists only of soil but at borehole SW 28, sited on alluvium, 1.8 m of soil and peaty silt were proved.

Block C (Table 5, Fig. 7)
This block includes the remainder of the First Terrace of the River Trent. Just north of Gainsborough boreholes SW 42 and 43 proved no potentially workable sand and gravel, and an area around these holes has been shown as mineral-free on the resource map. Elsewhere mineral consists of First Terrace deposits underlain in places by 25-Ft Drift sand and Older River Sand and Gravel.

Over most of the block, Older River Sand and Gravel is absent, but boreholes NW 21 and 23 and SW 35 found about 2.0 m, 1.2 m and 1.1 m respectively of this deposit consisting of pebbly sand with a mean gravel content of 12 per cent. Half the IMAU boreholes proved 25-Ft Drift sand ranging in thickness from 1.0 to 3.5 m and having a mean grading of 2 per cent fines, 97 per cent sand and 1 per cent gravel. The ubiquitous First Terrace deposits, with proved thicknesses from 1.1 to 7.0 m, generally consist of sand with only traces of gravel but, exceptionally, a sample from the base of the deposit in borehole NW 21 contained 40 per cent of pebbles.

Total mineral thicknesses proved in the boreholes range from 2.0 to 10.5 m, giving a mean of 5.5 m. The estimated volume of mineral is 56 million $m^3 \pm 41$ per cent and the mean grading is 5 per cent fines, 93 per cent sand and 2 per cent gravel.

Where both 25-Ft Drift and First Terrace mineral deposits occur the two are separated by a clay or silt parting ranging in thickness from 0.7 to 2.6 m. In addition at borehole NW 25, a 1.0-m clay parting is present in 25-Ft Drift sand. The mean thickness of waste in the block is 1.0 m.

Peat up to at least 1.5 m thick covers much of the block; elsewhere mineral lies beneath relatively thin soil or alluvial clays and silts. The mean thickness of overburden for the block is 0. 6 m.

The mineral of blocks C and D taken together has a mean thickness of 8.6 m and an estimated volume of 154 million $m^3 \pm 24$ per cent.

Block D (Table 6, Fig. 8)
The mineral within this block consists almost entirely of blown sand with a mean thickness of 2.5 m and a mean grading of 12 per cent fines and 88 per cent sand. The blown sand rests on Lias bedrock or on Boulder Clay and the small amounts of gravel found at the base may be a remanié deposit resulting from erosion of the latter. Near Scotton, borehole NE 116 proved 3.0 m of clayey sand, classified as alluvium, separated by 1.1 m of clayey silt from 1.5 m of sandy gravel of the Terrace of the River Eau, and the inclusion of these deposits results in a mean grading for the

Table 5. Block C: data from assessment boreholes

Borehole No.	Recorded thickness (m)			Mean grading percentage					
	Mineral	Overburden	Waste Partings	Fines	Fine Sand	Medium Sand	Coarse Sand	Fine Gravel	Coarse Gravel
				-1/16mm	$-\frac{1}{4}+1/16$mm	$-1+\frac{1}{4}$mm	-4+1mm	-16+4mm	+16mm
NW 21	10.0	0.4	1.6	7	51	35	1	1	5
23	8.2	0.7	1.6	8	55	34	1	1	1
25	10.5	0.5	1.7	4	49	46	1	trace	
26	2.6	0.5	1.9	5	64	23	4	3	1
28	2.0	0.5		2	27	70	1	trace	
SW 31	9.8	0.4	2.6	3	47	50	trace		
32	2.0	1.0		4	33	62	1	trace	
35	5.5	1.5	1.1	2	34	57	4	1	2
38	2.3	nil		2	20	77	1		
39	3.1	0.5		5	31	63	1		
40	4.0	0.4		6	14	79	1	trace	
41	5.7	0.3	1.8	4	30	65	1		
42	Absent								
43	(0.6)	(0.4)		27	12	58	2	1	

Table 6. Block D: data from assessment boreholes

Borehole No.	Recorded thickness (m)			Mean grading percentage					
	Mineral	Overburden	Waste Partings	Fines	Fine Sand	Medium Sand	Coarse Sand	Fine Gravel	Coarse Gravel
				-1/16mm	$-\frac{1}{4}+1/16$mm	$-1+\frac{1}{4}$mm	-4+1mm	-16+4mm	+16mm
NE 108	2.0	nil		9	31	59	1		
109	4.6	0.4		2	29	69	trace	trace	
111	2.0	0.3		5	34	61	trace		
112	1.7	nil		28	45	26	1	trace	
113	2.5	0.2		5	35	60	trace		
114	2.5	0.3		6	39	55	trace		
115	2.0	0.5		28	43	26	trace	1	2
116	4.5	0.4	1.1	16	26	35	8	12	3

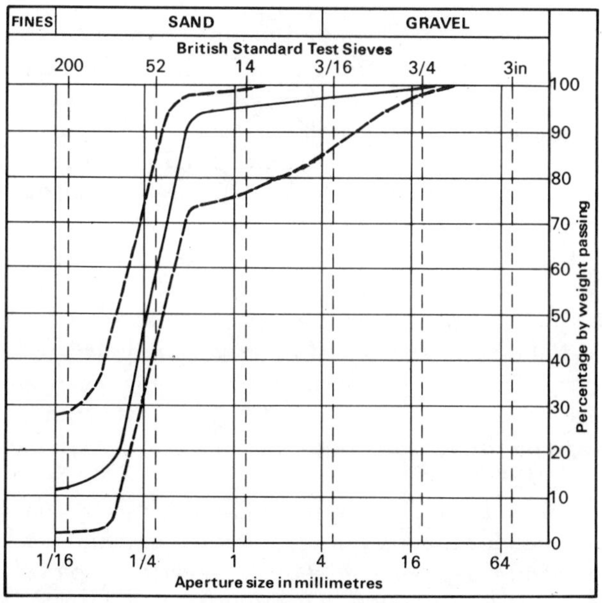

Figure 8. Grading characteristics of the mineral in block D (for explanation see Fig. 5)

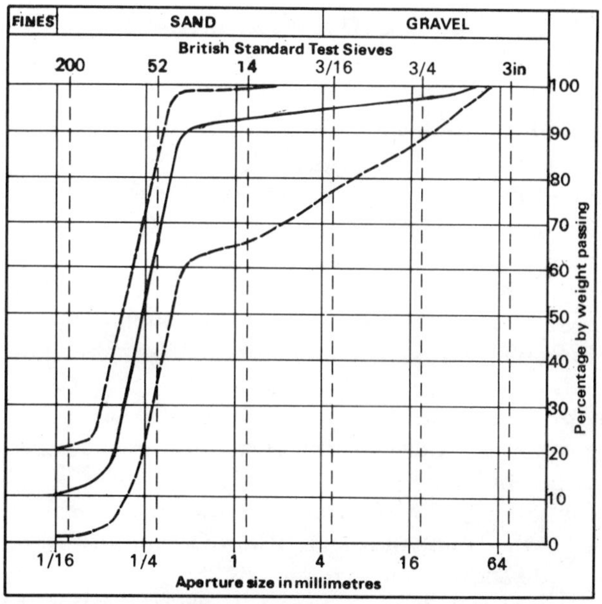

Figure 9. Grading characteristics of the mineral in block E (for explanation see Fig. 5)

block of 11 per cent fines, 86 per cent sand and 3 per cent gravel.

The estimated volume of mineral present is 26 million m^3 $\pm$ 37 per cent.

Block E (Table 7, Fig. 9)
The mineral in this block consists of Glacial Sand and Gravel in three separate areas. Around and north of Laughton the Glacial Sand and Gravel comprises pebbly sand and sandy gravel and is in part overlain by blown sand; in this area borehole NE 107 found no sand and gravel but field observations by R. J. Bull indicate the presence of sand up to at least one metre thick in close proximity to this site. Around Blyton the deposit ranges in composition from sand to 'very clayey' pebbly sand and sandy gravel and in proved thickness from 0.9 to 4.4 m; boreholes NW 29 and 31 included waste partings 2.0 m and 1.3 m thick respectively. In the Pilham—Corringham area the Glacial Sand and Gravel is in the form of a lens in Boulder Clay and its extent is uncertain; it consists of sand, commonly 'clayey', which ranges up to 5.0 m in thickness; in borehole SE 116 it included a 0.5-m clay parting.

The mean thickness of mineral in the block is 2.6 m and the mean grading is 10 per cent fines, 85 per cent sand and 5 per cent gravel. The estimated volume is 24 million m^3 $\pm$ 39 per cent.

North of Laughton and around Blyton overburden is thin but in the Pilham—Corringham area up to 4.9 m of soil and Boulder Clay have been proved above the sand.

Table 7. Block E: data from assessment boreholes

Borehole No.	Recorded thickness (m)			Mean grading percentage					
	Mineral	Overburden	Waste Partings	Fines	Fine Sand	Medium Sand	Coarse Sand	Fine Gravel	Coarse Gravel
				-1/16mm	$-\frac{1}{4}$+1/16mm	$-1+\frac{1}{4}$mm	-4+1mm	-16+4mm	+16mm
NW 29	2.1	nil	2.0	15	26	42	3	6	8
30	1.6	0.4		10	33	22	10	12	13
31	3.2	0.5	1.3	6	63	24	1	1	5
NE 107	Absent								
110	4.4	0.7		20	29	38	5	5	3
SW 44	(0.9)	0.4		7	22	68	1	2	
45	2.0	0.3		1	21	78	trace	trace	
SE 116	3.5	2.5	0.5	8	27	64	1	trace	
117	2.0	4.9		10	64	26	trace		
118	5.0	0.3		12	46	39	1	1	1
119	3.9	3.6		1	60	38	1	trace	
120	Absent								

APPENDIX A: FIELD AND LABORATORY PROCEDURES

Trial and error during initial studies of the complex and variable glacial deposits of East Anglia and Essex showed that an absolute minimum of five sample points evenly distributed across the sand and gravel are needed to provide a worthwhile statistical assessment, but that, where possible, there should be not less than ten. Sample points are any points for which adequate information exists about the nature and thickness of the deposit and may include boreholes other than those drilled during the survey and exposures. In particular, the cooperation of sand and gravel operators ensures that boreholes are not drilled where reliable information is already available; although this may be used in the calculations, it is held confidentially by the Institute and cannot be disclosed.

The mineral shown on each 1:25 000 sheet is divided into resource blocks. The arbitrary size selected, 10 km², is a compromise to meet the aims of the survey by providing sufficient sample points in each block. As far as possible the block boundaries are determined by geological boundaries so that, for example, glacial and river terrace gravels are separated. Otherwise division is by arbitrary lines, which may bear no relationship to the geology. The blocks are drawn provisionally before drilling begins.

A reconnaissance of the ground is carried out to record any exposures and inquiries are made to ascertain what borehole information is available. Borehole sites are then selected to provide an even pattern of sample points at a density of approximately one per square kilometre. However, because broad trends are independently overlain by smaller scale characteristically random variations, it is unnecessary to adhere to a square grid pattern. Thus such factors as ease of access and the need to minimise disturbance to land and the public are taken into account in siting the holes; at the same time it is necessary to guard against the possibility that ease of access (that is, the positions of roads and farms) may reflect particular geological conditions, which may bias the drilling results.

The drilling machine employed should be capable of providing a continuous sample representative of all unconsolidated deposits, so that the in-situ grading can be determined, if necessary, to a depth of 30 m (100 ft) at a diameter of about 200 mm (8 in), beneath different types of overburden. It should be reliable, quiet, mobile and relatively small (so that it can be moved to sites of difficult access). Shell and auger rigs have proved to be almost ideal.

The rigs are modified to enable deposits above the water table to be drilled 'dry', instead of with water added to facilitate the drilling, to minimise the amount of material drawn in from outside the limits of the hole. The samples thus obtained are representative of the in-situ grading, and satisfy one of the most important aims of the survey. Below the water table the rigs are used conventionally, although this may result in the loss of some of the fines fraction and the pumping action of the bailer tends to draw unwanted material into the hole from the sides or the bottom.

A continuous series of bulk samples is taken throughout the sand and gravel. Ideally samples are composed exclusively of the whole of the material encountered in the borehole between stated depths. However, care is taken to discard, as far as possible, material which has caved or has been pumped from the bottom of the hole. A new sample is commenced whenever there is an appreciable lithological change within the sand and gravel, or at every 1 m (3.3 ft) depth. The samples, each weighing between 25 and 45 kg (55 and 100 lb), are despatched in heavy duty polythene bags to a laboratory for grading. The grading procedure is based on British Standard 1377 (1967). Random checks on the accuracy of the grading are made in the Institute's laboratories.

All data, including mean grading analysis figures calculated for the total thickness of the mineral, are entered on standard record sheets, abbreviated copies of which are reproduced in Appendix F.

Detailed records may be consulted at the appropriate offices of the Institute, upon application to the Head, Industrial Minerals Assessment Unit.

APPENDIX B: STATISTICAL PROCEDURE

Statistical Assessment

1. A statistical assessment is made of an area of mineral greater than 2 km², if there is a minimum of five evenly spaced boreholes in the resource block (for smaller areas see paragraph 12 below).

2. The simple methods used in the calculations are consistent with the amount of data provided by the survey. Conventional symmetrical confidence limits are calculated for the 95 per cent probability level, that is, there is a 5 per cent or one in twenty chance of a result falling outside the stated limits.

3. The volume estimate (V) for the mineral in a given block is the product of the two variables, the sampled areas (A) and the mean thickness ($\bar{l}_m$) calculated from the individual thicknesses at the sample points. The standard deviations for these variables are related such that

$$S_V = \sqrt{(S_A{}^2 + S_{\bar{l}_m}{}^2)} \qquad [1]$$

4. The above relationship may be transposed such that

$$S_V = S_{\bar{l}_m} \sqrt{(1 + S_A{}^2/S_{\bar{l}_m}{}^2)} \qquad [2]$$

From this it can be seen that as $S_A{}^2/S_{\bar{l}_m}{}^2$ tends to 0, S_V tends to $S_{\bar{l}_m}$.
If, therefore, the standard deviation for area is small with respect to that for mean thickness, the standard deviation for volume approximates to that for mean thickness.

5. Given that the number of approximately evenly spaced sample points in the sampled area is n with mineral thickness measurements $l_{m_1}, l_{m_2}, \ldots l_{m_n}$, then the best estimate of mean thickness, $\bar{l}_m$, is given by

$$\frac{\sum(l_{m_1}+l_{m_2}\ldots l_{m_n})}{n}$$

For groups of closely spaced boreholes a discretionary weighting factor may be applied to avoid bias (see note on weighting below). The standard deviation for mean thickness, $S_{\bar{l}}$, expressed as a proportion of the mean thickness is given by

$$S_{\bar{l}}=(1/\bar{l}_m)\sqrt{[(l_m-\bar{l}_m)^2/(n-1)]}$$

where l_m is any value in the series l_{m_1} to l_{m_n}.

6. The sampled area in each resource block is coloured pink on the map. Wherever possible, calculations relate to the mineral within mapped geological boundaries (which may not necessarily correspond to the limits of deposit). Where the area is not defined by a mapped boundary, that is, where the boundary is inferred, a distinctive symbol is used. Experience suggests that the errors in determining area are small relative to those in thickness. The relationship $S_A/S_{\bar{l}_m} \leqslant 1/3$ is assumed in all cases. It follows from equation [2] that

$$S_{\bar{l}_m} \leqslant S_V \leqslant 1.05\, S_{\bar{l}_m} \qquad [3]$$

7. The limits on the estimate of mean thickness of mineral, $L_{\bar{l}_m}$, may be expressed in absolute units $\pm(t/\sqrt{n})\times S_{\bar{l}_m}$ or as a percentage $\pm(t/\sqrt{n})\times S_{\bar{l}_m}\times(100/\bar{l}_m)$ per cent, where t is Student's t at the 95 per cent probability level for $(n-1)$ degrees of freedom, evaluated by reference to statistical tables. (In applying Student's t it is assumed that the measurements are distributed normally).

8. Values of t at the 95 per cent probability level for values of n up to 20 are as follows:

n	t	*n*	t
1	infinity	11	2.228
2	12.706	12	2.201
3	4.303	13	2.179
4	3.182	14	2.160
5	2.776	15	2.145
6	2.571	16	2.131
7	2.447	17	2.120
8	2.365	18	2.110
9	2.306	19	2.101
10	2.262	20	2.093

(from Table 12, Biometrika Tables for Statisticians, Volume 1, Second Edition, Cambridge University Press, 1962). When *n* is greater than 20, 1.96 is used (the value of t when *n* is infinity).

9. In calculating confidence limits for volume, L_V, the following inequality corresponding to equation [3] is applied: $L_{\bar{l}_m} \leqslant L_V \leqslant 1.05\, L_{\bar{l}_m}$

10. In summary, for values of *n* between 5 and 20, L_V is calculated as

$$[(1.05\times t)/\bar{l}_m]\times[\sqrt{\Sigma(l_m-\bar{l}_m)^2/n(n-1)}]\times 100$$

per cent, and when *n* is greater than 20, as

$$[(1.05\times 1.96)/\bar{l}_m]\times[\sqrt{\Sigma(l_m-\bar{l}_m)^2/n(n-1)}]\times 100$$

per cent.

11. The application of this procedure to a fictitious area is illustrated in Figs. 10 and 11

12. If the sampled area of mineral in a resource block is between 0.25 km^2 and 2 km^2 an assessment is inferred, based on geological and topographical information usually supported by the data from one or two boreholes. The volume of mineral is calculated as the product of the area, measured from field data, and the estimated thickness. Confidence limits are not calculated.

13. In some cases a resource block may include an area left uncoloured on the map, within which mineral (as defined) is interpreted to be generally absent. If there is reason to believe that some mineral may be present, an inferred assessment may be made.

14. No assessment is attempted for an isolated area of mineral less than 0.25 km^2.

15. *Note on Weighting* The thickness of a deposit at any point may be governed solely by the position of the point in relation to a broad trend. However, most sand and gravel deposits also exhibit a random pattern of local, and sometimes considerable, variation in thickness. Thus the distribution of sample points need be only approximately regular and in estimating the mean thickness only simple weighting is necessary. In practice, equal weighting can often be applied to thicknesses at all sample points. If, however, there is a distinctly unequal distribution of points, bias is avoided by dividing the sampled area into broad zones, to each of which a value roughly proportional to its area is assigned. This value is then shared between the data points within the zone as the weighting factor.

APPENDIX C: CLASSIFICATION AND DESCRIPTION OF SAND AND GRAVEL

For the purposes of assessing resources of sand and gravel a classification should take account of economically important characteristics of the deposit, in particular the absolute content of fines and the ratio of sand to gravel.

The terminology commonly used by geologists when describing sedimentary rocks (Wentworth, 1922) is not entirely satisfactory for this purpose. For example, Wentworth proposed that a deposit should be described as a 'gravelly sand' when it contains more sand than gravel and there is at least 10 per cent of gravel, provided that there is less than 10 per cent of material finer than sand (less than 1/16 mm) and coarser than pebbles (more than 64 mm in diameter). Because deposits containing more than 10 per cent fines are not embraced by this system a modified binary classification based on Willman (1942) has been adopted.

When the fines content exceeds 40 per cent the material is not considered to be potentially workable and falls outside the definition of mineral. Deposits which contain 40 per cent fines or less are classified primarily on the ratio of sand to gravel but qualified in the light of the fines content, as follows: less than 10 per cent fines – no qualification; 10 per cent or more but less than 20 per cent fines – 'clayey'; 20 to 40 per cent fines – 'very clayey'.

The term 'clay' (as written, with single quote marks) is used to describe all material passing 1/16 mm. Thus it has no mineralogical significance and includes particles falling within the size range of silt. The normal meaning applies to the term clay where it does not appear in single quotation marks.

The ratio of sand to gravel defines the boundaries between sand, pebbly sand, sandy gravel and gravel (at 19:1, 3:1 and 1:1).

Thus it is possible to classify the mineral into one of twelve descriptive categories (see Fig.12). The procedure is as follows:

1. Classify according to ratio of sand to gravel.
2. Describe fines.

For example, a deposit grading 11 per cent gravel, 70 per cent sand and 19 per cent fines is classified as 'clayey' pebbly sand. This short description is included in the borehole log (see Note 9, p. 20).

Many differing proposals exist for the classification of the grain size of sediments (Atterberg, 1905; Udden, 1914; Wentworth, 1922; Wentworth, 1935; Allen, 1936; Twenhofel, 1937; Lane and others, 1947). As Archer (1970a, b) has emphasised, there is a pressing need for a simple metric scale acceptable to both scientific and engineering interests, for which the class limit sizes correspond closely with certain marked changes in the natural properties of mineral particles. For example, there is an important change in the degree of cohesion between particles at about the 1/16-mm size, which approximates to the generally accepted boundary between silt and sand. These and other requirements are met by a system based on Udden's geometric scale and a simplified form of Wentworth's terminology (Table 8), which is used in this Report.

The fairly wide intervals in the scale are consistent with the general level of accuracy of the qualitative assessments of the resource blocks. Three sizes of sand are recognised, fine (−¼ +1/16 mm), medium (−1 +¼ mm) and coarse (−4 +1 mm). The boundary at 16 mm distinguishes a range of finer gravel (−16 +4 mm), often characterised by abundance of worn tough pebbles of vein quartz, from larger pebbles often of notably different materials. The boundary at 64 mm distinguishes pebbles from cobbles. The term 'gravel' is used loosely to denote both pebble-sized and cobble-sized material.

The size distribution of borehole samples is determined by sieve analysis. In this report the grading is tabulated on the borehole record sheets (Appendix F), the intercepts corresponding with the simple geometric scale 1/16 mm, ¼ mm, 1 mm, 4 mm, 16 mm and so on as required. Original sample gradings are available for reference at the appropriate office of the Institute..

Each bulk sample is described, subjectively, by a geologist at the borehole site. Being based on visual examination, the description of the grading is inexact, the accuracy depending on the experience of the observer. The descriptions recorded are modified, as necessary, when the laboratory results become available.

The relative proportions of the rock types present in the gravel fraction are indicated by the use of the words 'and' or 'with'. For example, 'flint and quartz' indicates very approximate equal proportions with neither constituent accounting for less than about 25 per cent of the whole; 'flint with quartz' indicates that flint is dominant and quartz, the principal accessory rock types, comprises 5 to 25 per cent of the whole. Where the accessory material accounts for less than 5 per cent of the whole, but is still readily apparent, the phrase 'with some' has been used. Rare constituents are referred to as 'trace'.

The terms used in the field to describe the degree of rounding of particles, which is concerned with the sharpness of the edges and corners of a clastic fragment and not the shape (after Pettijohn, 1957), are as follows.

Angular: showing little or no evidence of wear; sharp edges and corners.

Subangular: showing definite effects of wear. Fragments still have their original form but edges and corners begin to be rounded off.

Subrounded: showing considerable wear. The edges and corners are rounded off to smooth curves. Original grain shape is still distinct.

Rounded: original faces almost completely destroyed, but some comparatively flat surfaces may still remain. All original edges and corners have been smoothed off to rather broad curves. Original shape is still apparent.

Well-rounded: no original faces, edges or corners left. The entire surface consists of broad curves; flat areas are absent. The original shape is suggested by the present form of the grain.

Table 8 Classification of gravel, sand and fines

Size limits	Grain size description	Qualification	Primary classification
	Cobble		Gravel
64 mm	Pebble	Coarse	
16 mm		Fine	
4 mm	Sand	Coarse	Sand
1 mm		Medium	
¼ mm		Fine	
1/16mm	Fines (silt and clay)		Fines

Block Calculation 1:25 000 Block } Fictitious

Area

Block: 11.08 km^2

Mineral: 8.32 km^2

Mean Thickness

Overburden: 2.5 m

Mineral: 6.5 m

Volume

Overburden: 21 million m^3

Mineral: 54 million m^3

Confidence limits of the estimate of mineral volume at the 95 per cent probability level: ± 20 per cent

That is, the volume of mineral (with 95 per cent probability): 54 ± 11 million m^3

Thickness estimate: measurements in metres

l_o = overburden thickness l_m = mineral thickness

Sample point	Weighting w	Overburden l_o	wl_o	Mineral l_m	wl_m	Remarks
SE 14	1	1.5	1.5	9.4	9.4	IMAU boreholes
SE 18	1	3.3	3.3	5.8	5.8	
SE 20	1	nil	-	6.9	6.9	
SE 22	1	0.7	0.7	6.4	6.4	
SE 23	1	6.2	6.2	4.1	4.1	
SE 24	1	4.3	4.3	6.4	6.4	
SE 17	½	1.2	1.6	9.8	7.2	Hydrogeological Dept record
123/45	½	2.0		4.6		
1	¼	2.7	2.6	7.3	5.8	Close group of four boreholes (commercial)
2	¼	4.5		3.2		
3	¼	0.4		6.8		
4	¼	2.8		5.9		
Totals	$\Sigma w = 8$	$\Sigma wl_o = 20.2$		$\Sigma wl_m = 52.0$		
Means		$l_o = 2.5$		$\bar{l}_m = 6.5$		

Calculation of confidence limits

l_m	$(l_m - \bar{l}_m)$	$(l_m - \bar{l}_m)^2$
9.4	2.9	8.41
5.8	0.7	0.49
6.9	0.4	0.16
6.4	0.1	0.01
4.1	2.4	5.76
6.4	0.1	0.01
7.2	0.7	0.49
5.8	0.7	0.49

$\Sigma(l_m - \bar{l}_m)^2 = 15.82$

$n = 8$

$t = 2.365$

L_V is calculated as

$$1.05 \times \frac{t}{\bar{l}_m} \sqrt{\frac{\Sigma(l_m - \bar{l}_m)^2}{n(n-1)}} \times 100$$

$$= 1.05 \times \frac{2.365}{6.5} \sqrt{\frac{15.82}{8 \times 7}} \times 100$$

$$= 20.3$$

$$\simeq 20 \text{ per cent}$$

Figure 10. Example of resource block assessment: calculations and results

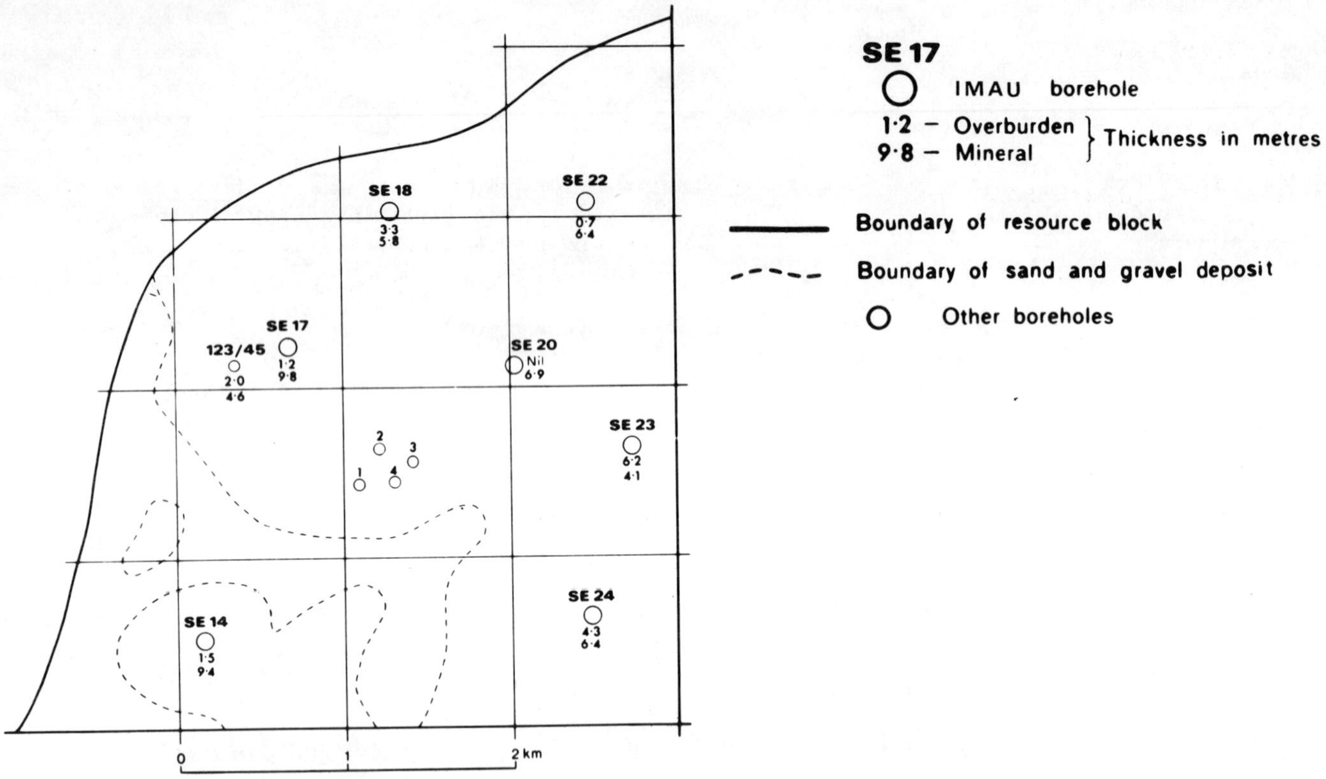

Figure 11. Example of resource block assessment: map of fictitious block

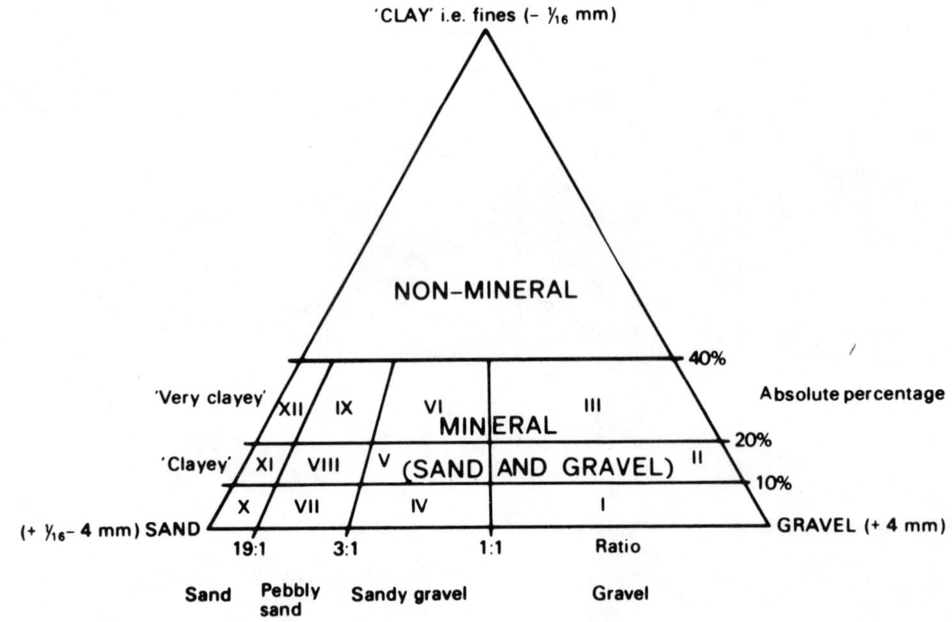

I	Gravel	IV	Sandy gravel	VII	Pebbly sand	X	Sand
II	'Clayey' gravel	V	Clayey sandy gravel	VIII	'Clayey' pebbly sand	XI	'Clayey' sand
III	'Very clayey' gravel	VI	'Very clayey' sandy gravel	IX	'Very clayey' pebbly sand	XII	'Very clayey' sand

Figure 12. Diagram to show the descriptive categories used in the classification of sand and gravel

APPENDIX D: EXPLANATION OF THE BOREHOLE RECORDS

Annotated Example

SK 89 SW 33[1] 8127 9400[2] Stockwith Road, East Stockwith Block B[3]

Surface level (+3.4 m) +11 ft[4]
Water level not recorded[5]
November 1972[6]

Overburden[7] 0.4 m
Mineral 6.0 m
Waste 0.8 m
Mineral 7.8 m
Bedrock 1.5 m+[8]

LOG

Geological Classification	Lithology	Thickness m	Depth m
	Soil	0.4	0.4
First Terrace	(a)[9] Sand with 0.2 m clay band at 2.4 m: medium, subangular to rounded quartz, quartzite and other lithic grains	6.0	6.4
25-Ft Drift	Clay, brown and grey	0.8	7.2
25-Ft Drift on Older River Sand and Gravel	(b) Sand, pebbly below 11.5 m; 0.1 m clay band at 8.4 m Gravel: fine, subangular to well rounded quartz and quartzite, subangular chert and flint and rounded sandstone, with some mudstone and siltstone Sand: medium, subangular to rounded quartz, quartzite and other lithic grains	7.8	15.0
Keuper Marl	Mudstone, red and grey	1.5+	16.5

GRADING

Mean for Deposit[12]

	%	mm	%
(a) Gravel	1	+ 16	0
		- 16 + 4	1
Sand	93	- 4 + 1	1
		- 1 + $\frac{1}{4}$	60
		- $\frac{1}{4}$ + 1/16	32
Fines	6	- 1/16	6
(b) Gravel	10	+ 16	2
		- 16 + 4	8
Sand	86	- 4 + 1	9
		- 1 + $\frac{1}{4}$	52
		- $\frac{1}{4}$ + 1/16	25
Fines	4	- 1/16	4

Bulk Samples

Depth below surface (m)	Percentage[11] Fines	Sand	Gravel
[10] 0.4 - 2.4	3	97	trace
2.4 - 2.6	clay band*		
2.6 - 5.6	4	96	trace
5.6 - 6.4	1	96	3
7.2 - 8.2	9	89	2
8.2 - 8.4	2	98	trace
8.4 - 8.5	clay band*		
8.5 - 9.5	1	98	1
9.5 - 10.5	2	94	4
10.5 - 11.5	3	96	1
11.5 - 12.5	1	76	23
12.5 - 13.5	2	81	17
13.5 - 14.5	trace	77	23
14.5 - 15.0	3	81	16

*Assumed to comprise 100% fines in calculating mean grading

COMPOSITION OF GRAVEL FRACTION

Percentages by weight (and number)

Depth below surface (m)	Quartzite	Quartz	Flint and chert	Sandstone	Mudstone and siltstone	Limestone
12.5 - 13.5	23(18)	23(27)	23(24)	20(22)	11(9)	trace
14.5 - 15.0	22(23)	39(26)	17(23)	12(13)	10(15)	-

The numbered paragraphs below correspond with the annotations given on the specimen record above.

1. Borehole Registration Number.
Each Industrial Minerals Assessment Unit (IMAU) borehole is identified by a Registration Number. This consists of two statements.

1) The number of the 1:25 000 sheet on which the borehole lies, for example SK 89.
2) The quarter of the 1:25 000 sheet on which the borehole lies and the number of the borehole in a series for that quarter, for example SW 33.

Thus the full Registration Number is SK 89 SW 33. Usually this is abbreviated to SW 33 in the text.

2. The National Grid Reference
All National Grid References in this publication lie within the 100-km square SK unless otherwise stated. Grid references are given to eight figures, accurate to within 10 m for borehole locations. (In the text, six-figure grid references are used for more approximate locations, for example, for farms).

3. Location
The position of the borehole is generally referred to the nearest named locality on the 1:25 000 base map and the resource block in which it lies is stated.

4. Surface level
The surface level at the borehole site is given in metres and feet above Ordnance Datum. All measurements were made in feet; approximate conversions to metres are given in brackets.

5. Groundwater conditions
Four kinds of entry are made: the record indicates the level at which groundwater stood on completion of drilling (in metres and feet above or below OD); or that water was encountered but its level not recorded; or that water was not encountered; or that no note of groundwater conditions was made.

6. Type of drill and date of drilling
Unless otherwise stated, all boreholes were drilled by a Dando shell and auger rig using 6-inch casing. The month and year of completion of the hole are stated.

7. Overburden, Mineral, Waste and Bedrock
Mineral is sand and gravel which, as part of a deposit, falls within the arbitrary definition of potentially workable material (see p. 1). Bedrock is the 'formation', 'country rock' or 'rock head' below which potentially workable sand and gravel will not be found. Waste is any material other than bedrock or mineral. Where waste occurs between the surface and mineral it is classified as overburden.

8. The plus sign (+) indicates that the base of the deposit was not reached during drilling.

9. Lithological Description
When sand and gravel is recorded a general description based on the grading characteristics (for details see Appendix C) is followed by more detailed particulars of the sand and/or gravel fraction. Where more than one mineral horizon is recognised each is designated by a letter e.g. (a), (b) etc. The description of other rocks is based on visual examination, in the field.

10. Sampling
A continuous series of bulk samples is taken throughout the thickness of sand and gravel. A new sample is commenced whenever there is an appreciable lithological change within the sand and gravel or at every 1 m of depth.

11. Grading Results
The limits are as follows: gravel, +4 mm; sand, -4+1/16 mm; fines, -1/16 mm.

12. Mean Grading
The grading of the full thickness of the mineral horizon identified in the log is the mean of the individual sample gradings weighted by the thicknesses represented, if these vary. The classification used is shown in Table 8.

Fully representative sampling of sand and gravel is difficult to achieve particularly where groundwater levels are high. Comparison between boreholes and adjacent exposures suggests that in borehole samples the proportion of sand may be higher and the proportions of fines and coarse gravel (+16 mm) may be lower.

APPENDIX E: LIST OF BOREHOLES USED IN THE ASSESSMENT OF RESOURCES

1. Industrial Minerals Assessment Unit Boreholes

Borehole No. by sheet quadrant	Grid reference (all fall in 100-km square SK)	Page No.
SK 89 NW		
4	8121 9963	22
5	8216 9947	23
6	8318 9945	24
7	8048 9862	25
8	8162 9865	26
9	8292 9889	27
10	8048 9766	28
11	8135 9767	29
12	8203 9819	30
13	8316 9824	31
14	8080 9683	32
15	8198 9714	33
16	8050 9615	34
17	8183 9603	35
18	8045 9550	36
19	8120 9526	37
20	8387 9868	38
21	8428 9760	39
22	8334 9702	40
23	8402 9678	41
24	8267 9629	42
25	8364 9610	43
26	8480 9639	44
27	8214 9545	45
28	8442 9558	46
29	8410 9938	47
30	8473 9805	48
31	8498 9513	49
SK 89 NE		
107	8511 9664	50
108	8505 9926	50
109	8541 9781	51
110	8572 9569	52
111	8606 9958	53

Borehole No.	Grid reference	Page No.
SK 89 NE		
112	8624 9861	54
113	8694 9910	55
114	8733 9845	56
115	8801 9957	57
116	8953 9961	58
SK 89 SW		
27	8073 9469	59
28	8194 9466	60
29	8007 9368	61
30	8025 9050	62
31	8311 9496	63
32	8404 9484	64
33	8127 9400	65-66
34	8259 9418	67
35	8355 9395	68
36	8070 9337	69
37	8206 9332	70
38	8311 9291	71
39	8418 9324	72
40	8032 9209	73
41	8124 9260	74
42	8227 9252	75
43	8131 9248	75
44	8498 9434	76
45	8430 9385	77
SK 89 SE		
116	8711 9365	78
117	8642 9268	79
118	8744 9254	80
119	8640 9154	81
120	8739 9104	82

2. Other boreholes

Numerous geophysical test holes indicate the absence of mineral from much of the eastern area.

APPENDIX F: INDUSTRIAL MINERALS ASSESSMENT UNIT BOREHOLE RECORDS

SK 89 NW 4 | 8121 9963 | Owston Ferry | Block A

Surface level (+3.7 m) +12 ft
Water level not recorded
September 1972

Waste 8.9 m
Bedrock 1.6 m+

LOG

Geological Classification	Lithology	Thickness m	Depth m
	Soil	0.5	0.5
Alluvium	Clay, brown and grey with 0.5 m brown peat at 3.0 m and 6.5 m	8.4	8.9
Keuper Marl	Mudstone, hard, green	1.6+	10.5

SK 89 NW 5	8216 9947	East Ferry	Block A

Surface level (+3.4 m) +11 ft
Water level (+1.2 m) +4 ft
January 1973

Overburden 6.0 m
Mineral 11.7 m
Bedrock 1.3 m+

LOG

Geological Classification	Lithology	Thickness m	Depth m
	Soil	0.2	0.2
Alluvium	Silt, light brown and grey, laminated	5.8	6.0
	(a) Sand, 'very clayey' at top: fine to medium subangular to rounded quartz with rock fragments	7.0	13.0
Older River Sand and Gravel	(b) Sandy gravel Gravel: coarse, subangular to well rounded quartz and quartzite with sandstone, and angular chert and flint Sand: medium, subangular to rounded quartz, quartzite and other rock fragments	4.7	17.7
Keuper Marl	Mudstone, red and grey	1.3+	19.0

GRADING

Mean for Deposit

	%	mm	%
(a) Gravel	1	+ 16	0
		- 16 + 4	1
Sand	90	- 4 + 1	3
		- 1 + ¼	46
		- ¼ + 1/16	41
Fines	9	- 1/16	9
(b) Gravel	34	+ 16	8
		- 16 + 4	26
Sand	65	- 4 + 1	8
		- 1 + ¼	47
		- ¼ + 1/16	10
Fines	1	- 1/16	1

Bulk Samples

Depth below surface (m)	Percentage Fines	Sand	Gravel
6.0 - 7.0	23	77	0
7.0 - 8.0	17	83	trace
8.0 - 9.0	7	89	4
9.0 - 10.0	6	93	1
10.0 - 11.0	5	93	2
11.0 - 12.0	2	96	2
12.0 - 13.0	2	97	1
13.0 - 14.0	2	54	44
14.0 - 15.0	1	52	47
15.0 - 16.0	2	74	24
16.0 - 17.0	1	72	27
17.0 - 17.7	2	75	23

SK 89 NW 6 | 8318 9945 | Pin Hill, East Ferry | Block A

Surface level (+2.7 m) +9 ft
Water level (+0.7 m) +2 ft
November 1972

Mineral 7.0 m
Waste 1.1 m
Mineral 3.9 m
Bedrock 1.5 m+

LOG

Geological Classification	Lithology	Thickness m	Depth m
First Terrace	(a) Sand, 'very clayey' at top: fine to medium subangular to rounded quartz, quartzite and rock fragments	7.0	7.0
25-Ft Drift	Clay, laminated, grey and brown	1.1	8.1
? 25-Ft Drift	(b) Sand, pebbly at base Gravel: fine, quartzite and quartz, with chert and sandstone and some mudstone and siltstone Sand: fine to medium, subangular to rounded quartz, quartzite and other lithic grains	3.9	12.0
Keuper Marl	Mudstone, red	1.5+	13.5

GRADING

Mean for Deposit

	%	mm	%
(a)			
Gravel	trace	+ 16	0
		- 16 + 4	trace
Sand	94	- 4 + 1	trace
		- 1 + $\frac{1}{4}$	49
		- $\frac{1}{4}$ + 1/16	45
Fines	6	- 1/16	6
(b)			
Gravel	2	+ 16	trace
		- 16 + 4	2
Sand	95	- 4 + 1	4
		- 1 + $\frac{1}{4}$	52
		- $\frac{1}{4}$ + 1/16	39
Fines	3	- 1/16	3

Bulk Samples

Depth below surface (m)	Percentage Fines	Sand	Gravel
0 - 0.4	35	65	trace
0.4 - 1.4	3	97	trace
1.4 - 2.4	13	87	trace
2.4 - 3.4	1	99	0
3.4 - 4.4	1	99	0
4.4 - 5.4	4	95	1
5.4 - 6.4	1	99	trace
6.4 - 7.0	2	98	trace
8.1 - 9.1	6	94	0
9.1 - 10.1	3	96	1
10.1 - 11.1	3	94	3
11.1 - 12.0	1	93	6

COMPOSITION OF GRAVEL FRACTION
Percentages by weight (and number)

Depth below surface (m)	Quartzite	Quartz	Flint and chert	Sandstone	Mudstone and siltstone	Limestone	Ironstone
11.1 - 12.0	27(26)	27(19)	17(18)	17(18)	10(14)	1(3)	1(2)

SK 89 NW 7 | 8048 9862 | Lady Croft Farm, Owston Ferry | Block A

Surface level (+4.3 m) +14 ft
Water not encountered
September 1972

Overburden 1.3 m
Mineral 1.0 m
Bedrock 1.2 m+

LOG

Geological Classification	Lithology	Thickness m	Depth m
	Soil	0.3	0.3
Alluvium	Clay, brown, with thin peat parting	1.0	1.3
	'Clayey' sand: fine, subangular to rounded quartz and quartzite	1.0	2.3
Keuper Marl	Mudstone, red and grey	1.2+	3.5

GRADING

Mean for Deposit

	%	mm	%
Gravel	1	+ 16	0
		- 16 + 4	1
		- 4 + 1	2
Sand	81	- 1 + $\frac{1}{4}$	33
		- $\frac{1}{4}$ + 1/16	46
Fines	18	- 1/16	18

Bulk Samples

Depth below surface (m)	Percentage Fines	Sand	Gravel
1.3 - 2.3	18	81	1

SK 89 NW 8	8162 9865	East Ferry	Block A

Surface level (+3.7 m) +12 ft
Water level (+0.6 m) +2 ft
January 1973

Overburden 11.3 m
Mineral 3.8 m
Bedrock 1.4 m+

LOG

Geological Classification	Lithology	Thickness m	Depth m
	Soil	0.3	0.3
Alluvium	Silt, light grey, with dark brown peat	11.0	11.3
Older River Sand and Gravel	Pebbly sand Gravel: fine, subrounded to well rounded quartz and quartzite with subangular chert Sand: medium, subangular to rounded quartz and quartzite with angular chert	3.8	15.1
Keuper Marl	Mudstone, red and grey	1.4+	16.5

GRADING

Mean for Deposit

	%	mm	%
Gravel	9	+ 16	2
		- 16 + 4	7
Sand	87	- 4 + 1	4
		- 1 + ¼	73
		- ¼ + 1/16	10
Fines	4	- 1/16	4

Bulk Samples

Depth below surface (m)	Percentage Fines	Sand	Gravel
11.3 - 12.3	3	85	12
12.3 - 13.3	3	85	12
13.3 - 14.3	4	93	3
14.3 - 15.1	5	87	8

SK 89 NW 9	8292 9889	Red Gate, East Ferry	Block A

Surface level (+2.1 m) +7.0 ft
Water level (-2.1 m) -7.0 ft
January 1973

Waste 15.6 m
Bedrock 1.4 m+

LOG

Geological Classification	Lithology	Thickness m	Depth m
	Soil	0.3	0.3
Alluvium	Silt, light brown and grey, mainly peaty	11.9	12.2
Older River Sand and Gravel	Pebbly sand Gravel: fine to coarse, subangular to rounded quartz and quartzite with some sandstone and angular to rounded chert Sand: medium, subangular to rounded quartz, quartzite and chert	3.4	15.6
Keuper Marl	Mudstone, red	1.4+	17.0

GRADING

Mean for Deposit

	%	mm	%
Gravel	23	+ 16	5
		- 16 + 4	18
Sand	71	- 4 + 1	5
		- 1 + $\frac{1}{4}$	33
		- $\frac{1}{4}$ + 1/16	33
Fines	6	- 1/16	6

Bulk Samples

Depth below surface (m)	Percentage Fines	Sand	Gravel
12.2 - 13.2	9	91	trace
13.2 - 14.2	5	74	21
14.2 - 15.2	4	48	48
15.2 - 15.6	5	72	23

SK 89 NW 10	8048 9766	Gunthorpe	Block A

Surface level (+2.4 m) +8 ft
Water level (+0.3 m) +1 ft
September 1972

Waste 9.0 m
Bedrock 2.0 m+

LOG

Geological Classification	Lithology	Thickness m	Depth m
	Soil	0.3	0.3
Alluvium	Clay and peat: 2.4 m of brown clay on 5.3 m peat with clay	7.7	8.0
	'Clayey' pebbly sand Gravel: fine to coarse, subrounded quartzite with quartz Sand: medium, subangular to subrounded quartz and quartzite	1.0	9.0
Keuper Marl	Mudstone, greyish green	2.0+	11.0

GRADING

Mean for Deposit

	%	mm	%
Gravel	6	+ 16	2
		- 16 + 4	4
Sand	78	- 4 + 1	2
		- 1 + ¼	45
		- ¼ + 1/16	31
Fines	16	- 1/16	16

Bulk Samples

Depth below surface (m)	Percentage Fines	Sand	Gravel
8.0 - 9.0	16	78	6

SK 89 NW 11	8135 9767	Wildsworth	Block A

Surface level (+3.0 m) +10 ft
Water level not recorded
January 1973

Overburden 10.0 m
Mineral 7.5 m
Bedrock 1.5 m+

LOG

Geological Classification	Lithology	Thickness m	Depth m
	Soil	0.3	0.3
Alluvium	Peaty silt: brown and black peat disseminated in grey silt	9.7	10.0
Older River Sand and Gravel	Sandy gravel Gravel: fine to coarse, subangular to well rounded quartz and quartzite with some sandstone and subangular chert Sand: medium subangular to rounded quartz and quartzite and other lithic grains	7.5	17.5
Keuper Marl	Mudstone, red	1.5+	19.0

GRADING

Mean for Deposit

	%	mm	%
Gravel	35	+ 16	9
		- 16 + 4	26
		- 4 + 1	9
Sand	64	- 1 + ¼	46
		- ¼ + 1/16	9
Fines	1	- 1/16	1

Bulk Samples

Depth below surface (m)	Percentage Fines	Sand	Gravel
10.0 - 11.0	3	55	42
11.0 - 12.0	1	85	14
12.0 - 13.0	1	75	24
13.0 - 14.0	2	80	18
14.0 - 15.0	1	32	67
15.0 - 16.0	1	60	39
16.0 - 17.0	1	71	28
17.0 - 17.5	1	36	63

SK 89 NW 12 | 8203 9819 | Wildsworth | Block A

Surface level (+2.4 m) +8 ft
Water level (+0.5 m) +2 ft
January 1973

Overburden 12.2 m
Mineral 4.8 m
Bedrock 3.0 m+

LOG

Geological Classification	Lithology	Thickness m	Depth m
	Soil	0.5	0.5
Alluvium	Peaty silt: 0.8 m brown silt on 10.9 m dark brown peaty silt	11.7	12.2
Older River Sand and Gravel	Sandy gravel, sandier at top Gravel: fine to coarse, subangular to well rounded quartz, quartzite and some sandstone, with angular chert and flint Sand: medium, subangular to rounded quartz, quartzite with chert and other lithic grains	4.8	17.0
Keuper Marl	Mudstone, red and grey	3.0+	20.0

GRADING

Mean for Deposit

	%	mm	%
Gravel	46	+ 16	12
		- 16 + 4	34
Sand	53	- 4 + 1	11
		- 1 + ¼	33
		- ¼ + 1/16	9
Fines	1	- 1/16	1

Bulk Samples

Depth below surface (m)	Percentage Fines	Sand	Gravel
12.2 - 13.2	2	74	24
13.2 - 14.2	2	52	46
14.2 - 15.2	2	56	42
15.2 - 16.2	1	33	66
16.2 - 17.0	1	54	45

SK 89 NW 13	8316 9824	Hornsey Hill, East Ferry	Block A

Surface level (+2.4 m) +8 ft
Water level (-0.7 m) -2 ft
January 1973

Overburden 1.4 m
Mineral 11.6 m
Bedrock 1.5 m+

LOG

Geological Classification	Lithology	Thickness m	Depth m
	Soil	0.4	0.4
Alluvium	Peaty silt, brown	1.0	1.4
First Terrace ? on 25-Ft Drift	(a) Sand: fine to medium, subangular to rounded quartz, quartzite and angular chert	10.0	11.4
Older River Sand and Gravel	(b) Sandy gravel Gravel: fine to coarse, subangular to well rounded quartz and quartzite with angular chert and some rounded sandstone Sand: medium, subangular to rounded quartz, quartzite and rock fragments	1.6	13.0
Keuper Marl	Mudstone, red	1.5+	14.5

GRADING

Mean for Deposit

	%	mm	%
(a)			
Gravel	1	+ 16	0
		- 16 + 4	1
		- 4 + 1	1
Sand	98	- 1 + $\frac{1}{4}$	51
		- $\frac{1}{4}$ + 1/16	46
Fines	1	- 1/16	1
(b)			
Gravel	35	+ 16	9
		- 16 + 4	26
		- 4 + 1	8
Sand	64	- 1 + $\frac{1}{4}$	38
		- $\frac{1}{4}$ + 1/16	18
Fines	1	- 1/16	1

Bulk Samples

Depth below surface (m)	Percentage Fines	Sand	Gravel
1.4 - 2.4	4	96	trace
2.4 - 3.4	4	96	trace
3.4 - 4.4	1	99	trace
4.4 - 5.4	1	99	0
5.4 - 6.4	1	99	trace
6.4 - 7.4	1	99	0
7.4 - 8.4	1	98	1
8.4 - 9.4	1	98	1
9.4 - 10.4	trace	99	1
10.4 - 11.4	1	97	2
11.4 - 12.4	1	59	40
12.4 - 13.0	1	74	25

SK 89 NW 14 | 8080 9683 | Wildsworth | Block A

Surface level (+3.0 m) +10 ft
Water level (+0.1 m) OD
January 1973

Overburden 11.7 m
Mineral 4.1 m
Bedrock 1.7 m+

LOG

Geological Classification	Lithology	Thickness m	Depth m
Alluvium	Peat and silt: 4.0 m brown and grey silt on brown, slightly silty peat	11.7	11.7
Older River Sand and Gravel	Pebbly sand, more gravelly at top Gravel: fine to coarse, subangular to well rounded quartz and quartzite, with subangular chert and some rounded sandstone and siltstone Sand: medium, subangular to rounded quartz, quartzite and other lithic grains	4.1	15.8
Keuper Marl	Mudstone, red	1.7+	17.5

GRADING

Mean for Deposit

	%	mm	%
Gravel	27	+16	8
		-16 + 4	19
Sand	72	- 4 + 1	15
		- 1 + $\frac{1}{4}$	47
		- $\frac{1}{4}$ + 1/16	10
Fines	1	- 1/16	1

Bulk Samples

Depth below surface (m)	Percentages Fines	Sand	Gravel
11.7 - 12.7	1	47	52
12.7 - 13.7	1	75	24
13.7 - 14.7	2	83	15
14.7 - 15.8	1	81	18

SK 89 NW 15	8198 9714	Bunkers Hill, Wildsworth	Block A

Surface level (+1.8 m) +6 ft
Water level (+0.7 m) +2 ft
January 1973

Overburden 10.4 m
Mineral 4.6 m
Bedrock 1.5 m+

LOG

Geological Classification	Lithology	Thickness m	Depth m
	Soil	0.4	0.4
Alluvium	Peaty silt: dark brown peat in varying concentrations within grey and brown silt	10.0	10.4
Older River Sand and Gravel	(a) Gravel, sandy in upper part Gravel: fine to coarse, subrounded to well rounded quartz and quartzite with angular chert and some rounded sandstone Sand: medium, subangular to rounded quartz, quartzite and other lithic grains	2.0	12.4
	(b) Pebbly sand Gravel: fine to coarse, well rounded quartz and quartzite with subangular chert Sand: medium, subangular to rounded quartz, quartzite and other lithic grains	2.6	15.0
Keuper Marl	Mudstone, red	1.5+	16.5

GRADING

Mean for Deposit

	%	mm	%
(a) Gravel	49	+ 16	8
		- 16 + 4	41
Sand	49	- 4 + 1	9
		- 1 + ¼	33
		- ¼ + 1/16	7
Fines	2	- 1/16	2
(b) Gravel	15	+ 16	6
		- 16 + 4	9
Sand	84	- 4 + 1	5
		- 1 + ¼	61
		- ¼ + 1/16	18
Fines	1	- 1/16	1

Bulk Samples

Depth below surface (m)	Percentage Fines	Sand	Gravel
10.4 - 11.4	3	59	38
11.4 - 12.4	1	38	61
12.4 - 13.4	2	84	14
13.4 - 14.4	1	85	14
14.4 - 15.0	1	83	16

SK 89 NW 16 8050 9615 Ravensfleet, Wildsworth Block A

Surface level (+2.1 m) +7 ft
Water level not recorded
January 1973

Overburden 7.0 m
Mineral 7.9 m
Bedrock 1.6 m+

LOG

Geological Classification	Lithology	Thickness m	Depth m
	Soil	0.2	0.2
Alluvium	Peaty silt, brown and black	6.8	7.0
	(a) Pebbly sand Gravel: fine to coarse, subangular to well rounded quartz and quartzite with subangular chert and some rounded sandstone and siltstone Sand: medium, subangular to rounded quartz, quartzite and rock fragments	4.0	11.0
Older River Sand and Gravel	(b) Sandy gravel Gravel: fine to coarse, subangular to well rounded quartz and quartzite with subangular chert and some rounded sandstone Sand: medium, subangular to rounded quartz and rock fragments	3.9	14.9
Keuper Marl	Mudstone, red	1.6+	16.5

GRADING

Mean for Deposit

	%	mm	%
(a) Gravel	14	+ 16	6
		- 16 + 4	8
		- 4 + 1	6
Sand	85	- 1 + $\frac{1}{4}$	68
		- $\frac{1}{4}$ + 1/16	11
Fines	1	- 1/16	1
(b) Gravel	44	+ 16	19
		- 16 + 4	25
		- 4 + 1	9
Sand	55	- 1 + $\frac{1}{4}$	39
		- $\frac{1}{4}$ + 1/16	7
Fines	1	- 1/16	1

Bulk Samples

Depth below surface (m)	Percentage Fines	Sand	Gravel
7.0 - 8.0	1	79	20
8.0 - 9.0	1	90	9
9.0 - 10.0	1	85	14
10.0 - 11.0	1	87	12
11.0 - 12.0	1	53	46
12.0 - 13.0	1	54	45
13.0 - 14.0	trace	66	34
14.0 - 14.9	1	45	54

SK 89 NW 17 8183 9603 Whoofer, Wildsworth Block A

Surface level (+1.8 m) +6 ft
Water level (-1.1 m) -4 ft
January 1973

Overburden 0.4 m
Mineral 5.0 m
Waste 0.6 m
Mineral 9.0 m
Bedrock 1.5 m+

LOG

Geological Classification	Lithology	Thickness m	Depth m
	Soil	0.4	0.4
First Terrace	(a) Sand, 'clayey' at top: fine to medium, subangular to rounded quartz, quartzite and other lithic grains	5.0	5.4
25-Ft Drift	Silt, grey and brown, laminated	0.6	6.0
25-Ft Drift on Older River Sand and Gravel	(b) Pebbly sand, 'clayey' at top Gravel: fine to coarse, subangular to rounded quartz and quartzite, with angular chert Sand: medium, subrounded quartz, quartzite and rock fragments	9.0	15.0
Keuper Marl	Mudstone, red	1.5+	16.5

GRADING

Mean for Deposit

	%	mm	%
(a)			
Gravel		+ 16	0
	trace	- 16 + 4	trace
		- 4 + 1	1
Sand	96	- 1 + 1/4	51
		- 1/4 + 1/16	44
Fines	4	- 1/16	4
(b)			
Gravel	9	+ 16	2
		- 16 + 4	7
		- 4 + 1	5
Sand	87	- 1 + 1/4	48
		- 1/4 + 1/16	34
Fines	4	- 1/16	4

Bulk Samples

Depth below surface (m)	Percentage Fines	Sand	Gravel
0.4 - 1.4	13	87	trace
1.4 - 2.4	2	98	trace
2.4 - 3.4	1	99	trace
3.4 - 4.4	1	99	trace
4.4 - 5.4	1	99	0
6.0 - 7.0	17	83	trace
7.0 - 8.0	3	94	3
8.0 - 9.0	1	75	24
9.0 - 10.0	4	85	11
10.0 - 11.0	3	92	5
11.0 - 12.0	5	80	15
12.0 - 13.0	3	87	10
13.0 - 14.0	2	89	9
14.0 - 15.0	3	89	8

SK 89 NW 18 | 8045 9550 | Greenhill Farm, Blyton | Block A

Surface level (+1.8 m) +6 ft
Water level (+0.7 m) +2 ft
November 1972

Overburden 6.4 m
Mineral 6.4 m
Bedrock 1.2 m+

LOG

Geological Classification	Lithology	Thickness m	Depth m
	Soil	0.4	0.4
Alluvium	Silt and peat: 1.3 m pale brown silt on 4.7 m brown peat	6.0	6.4
	(a) Pebbly sand Gravel: fine to coarse, subangular to well rounded quartz and quartzite with angular chert Sand: medium, angular to rounded quartz, quartzite and rock fragments	3.0	9.4
Older River Sand and Gravel	(b) Sandy gravel Gravel: fine to coarse, rounded to well rounded quartz and quartzite with subangular chert Sand: medium, angular to rounded quartz and rock fragments	3.4	12.8
Keuper Marl	Mudstone, red and pale grey	1.2+	14.0

GRADING

Mean for Deposit

	%	mm	%
(a) Gravel	7	+ 16	2
		- 16 + 4	5
Sand	92	- 4 + 1	7
		- 1 + ¼	71
		- ¼ + 1/16	14
Fines	1	- 1/16	1
(b) Gravel	42	+ 16	10
		- 16 + 4	32
Sand	57	- 4 + 1	11
		- 1 + ¼	38
		- ¼ + 1/16	8
Fines	1	- 1/16	1

Bulk Samples

Depth below surface (m)	Percentage Fines	Sand	Gravel
6.4 - 7.4	2	89	9
7.4 - 8.4	1	94	5
8.4 - 9.4	1	91	8
9.4 - 10.4	trace	52	48
10.4 - 11.4	1	51	48
11.4 - 12.4	1	65	34
12.4 - 12.8	1	58	41

SK 89 NW 19 | 8120 9526 | Greenhill Farm, Blyton | Block A

Surface level (+2.7 m) +9 ft
Water level (-0.1 m) OD
December 1972

Overburden 0.6 m
Mineral 9.4 m
Waste 0.5 m
Mineral 6.5 m
Bedrock 1.0 m+

LOG

Geological Classification	Lithology	Thickness m	Depth m
	Soil	0.6	0.6
First Terrace	(a) Sand: medium, subangular to well rounded quartz and rock fragments; coal specks below 6.0 m	9.4	10.0
25-Ft Drift	Silt, grey and brown	0.5	10.5
Older River Sand and Gravel	(b) Pebbly sand with gravel at base Gravel: fine to coarse, subrounded to well rounded quartz and quartzite, and subangular chert with rounded sandstone and some limestone and igneous rock Sand: medium, well rounded quartz and rock fragments	6.5	17.0
Keuper Marl	Mudstone, red and green	1.0+	18.0

GRADING

Mean for Deposit

	%	mm	%
(a)			
Gravel	trace	+ 16	0
		- 16 + 4	trace
Sand	97	- 4 + 1	trace
		- 1 + $\frac{1}{4}$	54
		- $\frac{1}{4}$ + 1/16	43
Fines	3	- 1/16	3
(b)			
Gravel	19	+ 16	7
		- 16 + 4	12
Sand	80	- 4 + 1	6
		- 1 + $\frac{1}{4}$	45
		- $\frac{1}{4}$ + 1/16	29
Fines	1	- 1/16	1

Bulk Samples

Depth below surface (m)	Percentage Fines	Sand	Gravel
0.6 - 1.6	4	96	trace
1.6 - 2.6	6	94	trace
2.6 - 3.6	1	99	trace
3.6 - 4.6	4	96	trace
4.6 - 5.6	1	99	trace
5.6 - 6.6	1	99	trace
6.6 - 7.6	1	99	trace
7.6 - 8.6	2	98	trace
8.6 - 9.6	4	96	trace
9.6 - 10.0	7	93	trace
10.5 - 11.5	1	90	9
11.5 - 12.5	2	91	7
12.5 - 13.5	2	93	5
13.5 - 14.5	1	94	5
14.5 - 15.5	2	84	14
15.5 - 17.0	1	44	55

COMPOSITION OF GRAVEL FRACTION
Percentages by weight (and number)

Depth below surface (m)	Quartzite	Quartz	Flint and chert	Sandstone	Mudstone and siltstone	Limestone
14.5 - 15.5	31(24)	24(28)	22(19)	16(19)	4(7)	3(3)

SK 89 NW 20 | 8387 9868 | Red Gate, East Ferry | Block B

Surface level (+4.6 m) +15 ft
Water level (+1.6 m) +5 ft
February 1973

Mineral 8.5 m
Waste 0.5 m
Mineral 4.4 m
Bedrock 1.6 m+

LOG

Geological Classification		Lithology	Thickness m	Depth m
First Terrace	(a)	Sand: fine to medium, subangular to rounded quartz and quartzite with subangular chert and rock fragments	8.5	8.5
25-Ft Drift		Silty clay, light brown and grey, laminated	0.5	9.0
? 25-Ft Drift	(b)	Sand: fine to medium, subrounded quartz and rock fragments; few fine rounded quartz and quartzite pebbles	4.4	13.4
Keuper Marl		Mudstone, red	1.6+	15.0

GRADING

Mean for Deposit

	%	mm	%
(a)			
Gravel	0	+ 16	0
		- 16 + 4	0
		- 4 + 1	1
Sand	98	- 1 + $\frac{1}{4}$	53
		- $\frac{1}{4}$ + 1/16	44
Fines	2	- 1/16	2
(b)			
Gravel	2	+ 16	1
		- 16 + 4	1
		- 4 + 1	2
Sand	95	- 1 + $\frac{1}{4}$	44
		- $\frac{1}{4}$ + 1/16	49
Fines	3	- 1/16	3

Bulk Samples

Depth below surface (m)	Percentage Fines	Sand	Gravel
0 - 1.0	4	96	0
1.0 - 2.0	2	98	0
2.0 - 3.0	2	98	0
3.0 - 4.0	2	98	0
4.0 - 5.0	2	98	0
5.0 - 6.0	1	99	0
6.0 - 7.0	1	99	0
7.0 - 8.0	1	99	trace
8.0 - 8.5	2	98	0
9.0 - 10.0	6	93	1
10.0 - 11.0	2	95	3
11.0 - 12.0	1	96	3
12.0 - 13.0	3	97	trace
13.0 - 13.4	2	95	3

SK 89 NW 21	8428 9760	Laughton	Block C

Surface level (+5.2 m) +17 ft
Water level not recorded
December 1972

Overburden 0.4 m
Mineral 7.0 m
Waste 1.6 m
Mineral 3.0 m
Bedrock 1.0 m+

LOG

Geological Classification	Lithology	Thickness m	Depth m
	Soil	0.4	0.4
First Terrace	(a) Sand, 'clayey' at base: fine to medium, subangular to rounded quartz, quartzite and rock fragments	6.0	6.4
	(b) 'Clayey' sand gravel Gravel: mainly coarse, subrounded to well rounded quartz and quartzite with subangular flint and chert Sand: fine, subrounded quartz, quartzite and rock fragments	1.0	7.4
25-Ft Drift	Clay, brown	1.6	9.0
Older River Sand and Gravel	(c) 'Clayey' pebbly sand Sand and Gravel as in (b)	3.0	12.0
Keuper Marl	Mudstone, green	1.0+	13.0

GRADING

Mean for Deposit

	%	mm	%
(a) & (b) Gravel	5	+ 16	4
		- 16 + 4	1
		- 4 + 1	trace
Sand	89	- 1 + $\frac{1}{4}$	43
		- $\frac{1}{4}$ + 1/16	46
Fines	6	- 1/16	6
(c) Gravel	7	+ 16	6
		- 16 + 4	1
		- 4 + 1	2
Sand	82	- 1 + $\frac{1}{4}$	16
		- $\frac{1}{4}$ + 1/16	64
Fines	11	- 1/16	11

Bulk Samples

Depth below surface (m)	Percentage Fines	Sand	Gravel
0.4 - 1.4	2	98	0
1.4 - 2.4	1	99	0
2.4 - 3.4	1	99	0
3.4 - 4.4	1	99	0
4.4 - 5.4	11	89	0
5.4 - 6.4	12	88	0
6.4 - 7.4	10	50	40
9.0 - 10.0	18	82	trace
10.0 - 11.0	4	75	21
11.0 - 12.0	10	90	trace

SK 89 NW 22 8334 9702 Mantree Hall, Laughton Block B

Surface level (+4.0 m) +13 ft
Water level (+1.0 m) +3 ft
January 1973

Overburden 0.3 m
Mineral 8.0 m
Waste 0.7 m
Mineral 5.0 m
Bedrock 1.5 m+

LOG

Geological Classification	Lithology	Thickness m	Depth m
	Soil	0.3	0.3
First Terrace	(a) Sand: fine to medium, subangular to well rounded quartz and quartzite with subangular chert	8.0	8.3
25-Ft Drift	Clay, brown, laminated	0.7	9.0
25-Ft Drift on Older River Sand and Gravel	(b) Sand, pebbly at base Gravel: fine, subangular to rounded quartz and quartzite, and some subangular chert Sand: fine to medium, subangular to rounded quartz and rock fragments	5.0	14.0
Keuper Marl	Mudstone, red	1.5+	15.5

GRADING

Mean for Deposit

	%	mm	%
(a)			
Gravel		+ 16	0
	trace	- 16 + 4	trace
		- 4 + 1	1
Sand	97	- 1 + ¼	56
		- ¼ + 1/16	40
Fines	3	- 1/16	3
(b)			
Gravel	2	+ 16	0
		- 16 + 4	2
Sand	95	- 4 + 1	3
		- 1 + ¼	48
		- ¼ + 1/16	44
Fines	3	- 1/16	3

Bulk Samples

Depth below surface (m)	Percentage Fines	Sand	Gravel
0.3 - 1.3	8	92	0
1.3 - 2.3	trace	100	0
2.3 - 3.3	1	99	0
3.3 - 4.3	1	99	0
4.3 - 5.3	1	99	trace
5.3 - 6.3	1	99	trace
6.3 - 7.3	7	93	trace
7.3 - 8.3	8	92	trace
9.0 - 10.0	8	92	trace
10.0 - 11.0	1	99	trace
11.0 - 12.0	2	98	trace
12.0 - 13.0	1	97	2
13.0 - 14.0	2	87	11

SK 89 NW 23 8402 9678 Laughton Block C

Surface level (+4.9 m) +16 ft
Water level (+1.7 m) +6 ft
January 1973

Overburden 0.7 m
Mineral 6.0 m
Waste 1.6 m
Mineral 2.2 m
Bedrock 1.5 m+

LOG

Geological Classification	Lithology	Thickness m	Depth m
Alluvium	Silt, brown	0.7	0.7
First Terrace	(a) Sand: fine to medium, subangular to rounded quartz, quartzite and other rock fragments	6.0	6.7
25-Ft Drift	Clay, brown, laminated	1.6	8.3
? 25-Ft Drift on Older River Sand and Gravel	(b) Pebbly sand with 0.2 m clay parting at 9.5 m Gravel: fine to coarse, angular to subrounded flint and chert with rounded quartz and quartzite Sand: fine to medium, subangular to rounded quartz and rock fragments	2.2	10.5
Keuper Marl	Mudstone, red	1.5+	12.0

GRADING

Mean for Deposit

	%	mm	%
(a)			
Gravel		+ 16	0
	trace	- 16 + 4	trace
		- 4 + 1	1
Sand	93	- 1 + ¼	34
		- ¼ + 1/16	58
Fines	7	- 1/16	7
(b)			
Gravel	7	+ 16	3
		- 16 + 4	4
		- 4 + 1	3
Sand	83	- 1 + ¼	35
		- ¼ + 1/16	45
Fines	10	- 1/16	10

Bulk Samples

Depth below surface (m)	Percentage Fines	Sand	Gravel
0.7 - 1.7	10	90	0
1.7 - 2.7	6	94	0
2.7 - 3.7	2	98	0
3.7 - 4.7	16	84	0
4.7 - 5.7	2	98	trace
5.7 - 6.7	5	95	trace
8.3 - 9.3	2	96	2
9.3 - 9.5	clay band*		
9.5 - 10.5	1	87	12

*Assumed to comprise 100% fines in calculating mean grading

SK 89 NW 24 8267 9629 Warp Farm, Laughton Block B

Surface level (+3.4 m) +11 ft
Water level not recorded
December 1972

Overburden 0.3 m
Mineral 14.9 m
Bedrock 1.3 m+

LOG

Geological Classification	Lithology	Thickness m	Depth m
	Soil	0.3	0.3
First Terrace	(a) Sand: medium, subrounded to well rounded quartz and rock fragments	9.0	9.3
? 25-Ft Drift on Older River Sand and Gravel	(b) Pebbly sand Gravel: fine, well rounded quartz, sandstone, limestone and chert Sand: as above with some subangular coal specks	5.9	15.2
Keuper Marl	Mudstone, greenish grey	1.3+	16.5

GRADING

Mean for Deposit

	%	mm	%
(a) Gravel	0	+ 16	0
		- 16 + 4	0
Sand	99	- 4 + 1	0
		- 1 + ¼	61
		- ¼ + 1/16	38
Fines	1	- 1/16	1
(b) Gravel	7	+ 16	2
		- 16 + 4	5
Sand	91	- 4 + 1	5
		- 1 + ¼	64
		- ¼ + 1/16	22
Fines	2	- 1/16	2

Bulk Samples

Depth below surface (m)	Percentage Fines	Sand	Gravel
0.3 - 1.3	2	98	0
1.3 - 2.3	1	99	0
2.3 - 3.3	1	99	0
3.3 - 4.3	1	99	trace
4.3 - 5.3	1	99	0
5.3 - 6.3	1	99	trace
6.3 - 7.3	1	99	trace
7.3 - 8.3	1	99	trace
8.3 - 9.3	1	99	trace
9.3 - 10.3	3	95	2
10.3 - 11.3	1	98	1
11.3 - 12.3	1	91	8
12.3 - 13.3	2	95	3
13.3 - 14.3	1	79	20
14.3 - 15.2	1	88	11

SK 89 NW 25   8364 9610   Pyewipe Hall, Laughton   Block C

Surface level (+3.0 m) +10 ft
Water level not recorded
January 1973

Overburden 0.5 m
Mineral 7.0 m
Waste 0.7 m
Mineral 1.5 m
Waste 1.0 m
Mineral 2.0 m
Bedrock 1.3 m+

LOG

Geological Classification	Lithology	Thickness m	Depth m
	Soil	0.5	0.5
First Terrace	(a) Sand: fine to medium, subangular to rounded quartz, quartzite and lithic grains	7.0	7.5
25-Ft Drift	Clay, brown	0.7	8.2
	(b) Sand, as above	1.5	9.7
	Clay, brown, several thin sand partings	1.0	10.7
	(c) Sand, as above	2.0	12.7
Keuper Marl	Mudstone, red	1.3+	14.0

GRADING

Mean for Deposit

	%	mm	%
(a)			
Gravel	trace	+ 16	0
		- 16 + 4	trace
Sand	96	- 4 + 1	1
		- 1 + $\frac{1}{4}$	48
		- $\frac{1}{4}$ + 1/16	47
Fines	4	- 1/16	4
(b)			
Gravel	trace	+ 16	0
		- 16 + 4	trace
Sand	94	- 4 + 1	trace
		- 1 + $\frac{1}{4}$	10
		- $\frac{1}{4}$ + 1/16	84
Fines	6	- 1/16	6
(c)			
Gravel	1	+ 16	0
		- 16 + 4	1
Sand	94	- 4 + 1	2
		- 1 + $\frac{1}{4}$	61
		- $\frac{1}{4}$ + 1/16	31
Fines	5	- 1/16	5

Bulk Samples

Depth below surface (m)	Percentage Fines	Sand	Gravel
0.5 - 1.5	7	93	0
1.5 - 2.5	10	90	0
2.5 - 3.5	7	93	0
3.5 - 4.5	1	99	trace
4.5 - 5.5	3	97	trace
5.5 - 6.5	1	99	0
6.5 - 7.5	trace	99	1
8.2 - 9.2	1	99	0
9.2 - 9.7	15	85	trace
10.7 - 11.7	6	94	trace
11.7 - 12.7	3	95	2

SK 89 NW 26	8480 9639	Laughton	Block C

Surface level (+3.0 m) +10 ft
Water level (+2.1 m) +7 ft
December 1972

Overburden 0.5 m
Mineral 1.1 m
Waste 1.9 m
Mineral 1.5 m
Bedrock 2.0 m+

LOG

Geological Classification	Lithology	Thickness m	Depth m
Alluvium	Clay, brown	0.5	0.5
First Terrace	(a) 'Clayey' sand: fine, subangular to well rounded quartz and quartzite with a few other lithic grains	1.1	1.6
25-Ft Drift	Silt, grey and brown, laminated	1.9	3.5
? Older River Sand and Gravel	(b) Sand, pebbly at top Gravel: fine to coarse, subrounded to well rounded quartz and quartzite with angular to subrounded chert and flint Sand: medium, subangular to rounded quartz, quartzite and other lithic grains	1.5	5.0
Keuper Marl	Mudstone, red	2.0+	7.0

GRADING

Mean for Deposit

	%	mm	%
(a) Gravel	2	+ 16	0
		- 16 + 4	2
Sand	87	- 4 + 1	4
		- 1 + ¼	31
		- ¼ + 1/16	52
Fines	11	- 1/16	11
(b) Gravel	5	+ 16	2
		- 16 + 4	3
Sand	94	- 4 + 1	3
		- 1 + ¼	17
		- ¼ + 1/16	74
Fines	1	- 1/16	1

Bulk Samples

Depth below surface (m)	Percentage Fines	Sand	Gravel
0.5 - 1.6	11	87	2
3.5 - 4.5	1	94	5
4.5 - 5.0	1	95	4

SK 89 NW 27 | 8214 9545 | Redhill Farm, Blyton | Block B

Surface level (+3.4 m) +11 ft
Water level (+1.7 m) +6 ft
December 1972

Overburden 0.4 m
Mineral 8.0 m
Waste 0.6 m
Mineral 5.2 m
Bedrock 1.3 m+

LOG

Geological Classification		Lithology	Thickness m	Depth m
		Soil	0.4	0.4
First Terrace	(a)	Sand, 'clayey' at top: fine to medium, subangular to rounded quartz, quartzite and other lithic grains	8.0	8.4
25-Ft Drift		Clay, brown, laminated	0.6	9.0
? 25-Ft Drift on Older River Sand and Gravel	(b)	Sand: medium, subrounded to well rounded quartz and rock fragments; well rounded quartz and quartzite pebbles in places.	5.2	14.2
Keuper Marl		Mudstone, green	1.3+	15.5

GRADING

Mean for Deposit

	%	mm	%
(a)			
Gravel	1	+ 16	1
		- 16 + 4	0
		- 4 + 1	1
Sand	95	- 1 + $\frac{1}{4}$	55
		- $\frac{1}{4}$ + 1/16	39
Fines	4	- 1/16	4
(b)			
Gravel	3	+ 16	1
		- 16 + 4	2
		- 4 + 1	4
Sand	96	- 1 + $\frac{1}{4}$	62
		- $\frac{1}{4}$ + 1/16	30
Fines	1	- 1/16	1

Bulk Samples

Depth below surface (m)	Percentage Fines	Sand	Gravel
0.4 - 1.4	11	89	0
1.4 - 2.4	6	94	0
2.4 - 3.4	6	94	trace
3.4 - 4.4	1	99	0
4.4 - 5.4	1	99	trace
5.4 - 6.4	5	95	0
6.4 - 7.4	2	98	0
7.4 - 8.4	2	95	3
9.0 - 10.0	1	98	1
10.0 - 11.0	1	97	2
11.0 - 12.0	1	96	3
12.0 - 13.0	1	87	12
13.0 - 14.2	1	97	2

SK 89 NW 28 | 8442 9558 | Sandbeck Farm, Blyton | Block C

Surface level (+5.2 m) +17 ft
Water level (+2.7 m) +9 ft
January 1973

Overburden 0.5 m
Mineral 2.0 m
Bedrock 2.5 m+

LOG

Geological Classification	Lithology	Thickness m	Depth m
	Soil	0.5	0.5
First Terrace	Sand: medium, subangular to well rounded quartz, quartzite, chert and other lithic grains	2.0	2.5
Keuper Marl	Mudstone, red	2.5+	5.0

GRADING

Mean for Deposit

	%	mm	%
Gravel		+ 16	0
	trace	- 16 + 4	trace
		- 4 + 1	2
Sand	99	- 1 + $\frac{1}{4}$	70
		- $\frac{1}{4}$ + 1/16	27
Fines	1	- 1/16	1

Bulk Samples

Depth below surface (m)	Percentage Fines	Sand	Gravel
0.5 - 1.5	1	99	0
1.5 - 2.5	2	97	1

SK 89 NW 29 | 8410 9938 | Hardwick Hill, East Ferry | Block E

Surface level (+34.1 m) +112 ft
Water not encountered
February 1973

Mineral 1.0 m
Waste 2.0 m
Mineral 1.1 m
Bedrock 1.4 m+

LOG

Geological Classification	Lithology	Thickness m	Depth m
Glacial Sand and Gravel	(a) Pebbly sand Gravel: fine to coarse, subangular to rounded quartzite and sandstone with siltstone and some subangular chert Sand: medium, angular to rounded quartz and quartzite with subangular chert and some subrounded lithic grains	1.0	1.0
Boulder Clay	Pebbly clay, hard, red	2.0	3.0
Glacial Sand and Gravel	(b) 'Very clayey' pebbly sand Gravel: mainly coarse, mudstone with quartz and quartzite Sand: medium, quartz, mudstone and other lithic grains	1.1	4.1
Keuper Marl	Mudstone, red, weathered	1.4+	5.5

GRADING

Mean for Deposit

%	mm	%
(a) & (b) Gravel 14	+ 16	8
	- 16 + 4	6
	- 4 + 1	3
Sand 71	- 1 + $\frac{1}{4}$	42
	- $\frac{1}{4}$ + 1/16	26
Fines 15	- 1/16	15

Bulk Samples

Depth below surface (m)	Percentage Fines	Sand	Gravel
0 - 1.0	8	79	13
3.0 - 4.1	21	64	15

SK 89 NW 30	8473 9805	Laughton	Block E

Surface level (+12.5 m) +41 ft
Water not encountered
December 1972

Overburden 0.4 m
Mineral 1.6 m
Bedrock 2.0 m+

LOG

Geological Classification	Lithology	Thickness m	Depth m
	Soil	0.4	0.4
Glacial Sand and Gravel	Sandy gravel, 'clayey' at base Gravel: fine to coarse, subrounded quartz and quartzite with subangular flint and chert Sand: medium, angular to subrounded quartz, quartzite and rock fragments	1.6	2.0
Keuper Marl	Mudstone, greenish grey	2.0+	4.0

GRADING

Mean for Deposit

	%	mm	%
Gravel	25	+ 16	13
		- 16 + 4	12
Sand	65	- 4 + 1	10
		- 1 + $\frac{1}{4}$	22
		- $\frac{1}{4}$ + 1/16	33
Fines	10	- 1/16	10

Bulk Samples

Depth below surface (m)	Percentage Fines	Sand	Gravel
0.4 - 1.4	8	65	27
1.4 - 2.0	14	63	23

SK 89 NW 31 | 8498 9513 | Blyton | Block E

Surface level (+21.0 m) +69 ft
Water level (+17.9 m) +59 ft
December 1972

Overburden 0.5 m
Mineral 1.2 m
Waste 1.3 m
Mineral 2.0 m
Bedrock 1.5 m+

LOG

Geological Classification	Lithology	Thickness m	Depth m
	Soil	0.5	0.5
Glacial Sand and Gravel	(a) 'Clayey' pebbly sand Gravel: fine to coarse, subangular to rounded quartz and quartzite with subangular chert Sand: fine to medium, subangular to well rounded quartz and quartzite with subrounded dark lithic grains	1.2	1.7
	Silt, pale grey and brown	1.3	3.0
	(b) Sand: as above but fine grained and with fewer dark lithic grains	2.0	5.0
Rhaetic	Mudstone, green	1.5+	6.5

GRADING

Mean for Deposit

	%	mm	%
(a)			
Gravel	16	+ 16	14
		- 16 + 4	2
		- 4 + 1	1
Sand	72	- 1 + ¼	32
		- ¼ + 1/16	39
Fines	12	- 1/16	12
(b)			
Gravel	0	+ 16	0
		- 16 + 4	0
		- 4 + 1	1
Sand	97	- 1 + ¼	19
		- ¼ + 1/16	77
Fines	3	- 1/16	3

Bulk Samples

Depth below surface (m)	Percentage Fines	Sand	Gravel
0.5 - 0.7	8	86	6
0.7 - 1.7	13	69	18
3.0 - 4.0	3	97	0
4.0 - 5.0	3	97	0

SK 89 NE 107 | 8511 9664 | Laughton | Block E

Surface level (+9.4 m) +31 ft
Water not encountered
December 1972

Waste 3.3 m
Bedrock 1.7 m+

LOG

Geological Classification	Lithology	Thickness m	Depth m
	Soil	0.4	0.4
Boulder Clay	Clay, greyish green	2.9	3.3
Rhaetic	Mudstone, greyish green	1.7+	5.0

SK 89 NE 108 | 8505 9926 | East Ferry | Block D

Surface level (+16.2 m) +53 ft
Water not encountered
February 1973

Mineral 2.0 m
Bedrock 2.0 m+

LOG

Geological Classification	Lithology	Thickness m	Depth m
Blown Sand	Sand: medium, subrounded quartz and quartzite with subangular chert and other lithic grains	2.0	2.0
Rhaetic	Mudstone, grey	2.0+	4.0

GRADING

Mean for Deposit

	%	mm	%
Gravel	0	+ 16	0
		- 16 + 4	0
Sand	91	- 4 + 1	1
		- 1 + ¼	59
		- ¼ + 1/16	31
Fines	9	- 1/16	9

Bulk Samples

Depth below surface (m)	Percentage Fines	Sand	Gravel
0.0 - 1.0	9	91	0
1.0 - 2.0	8	92	0

SK 89 NE 109 | 8541 9781 | Laughton | Block D

Surface level (+8.8 m) +29 ft
Water level (+6.8 m) +22 ft
December 1972

Overburden 0.4 m
Mineral 4.6 m
Waste 1.0 m
Bedrock 1.0 m+

LOG

Geological Classification	Lithology	Thickness m	Depth m
	Soil	0.4	0.4
Blown Sand	Sand: medium, subangular to well rounded quartz, quartzite and some lithic grains	4.6	5.0
? Boulder Clay	Silt, brown and grey	1.0	6.0
Rhaetic	Mudstone, grey	1.0+	7.0

GRADING

Mean for Deposit

	%	mm	%
Gravel	trace	+ 16	0
		- 16 + 4	trace
		- 4 + 1	trace
Sand	98	- 1 + $\frac{1}{4}$	69
		- $\frac{1}{4}$ + 1/16	29
Fines	2	- 1/16	2

Bulk Samples

Depth below surface (m)	Percentage Fines	Sand	Gravel
0.4 - 1.4	3	97	0
1.4 - 2.4	1	99	0
2.4 - 3.4	1	99	0
3.4 - 4.4	2	98	0
4.4 - 5.0	1	96	3

SK 89 NE 110	8572 9569	Blyton	Block E

Surface level (+15.5 m) +51 ft
Water level (+13.0 m) +43 ft
February 1973

Overburden 0.7 m
Mineral 4.4 m
Bedrock 1.4 m+

LOG

Geological Classification	Lithology	Thickness m	Depth m
	Soil	0.2	0.2
	Hardcore	0.5	0.7
Glacial Sand and Gravel	'Clayey' pebbly sand with 'clayey' sandy gravel at base Gravel: fine, subrounded quartz and quartzite with subangular chert and other lithic grains Sand: medium, subangular to rounded quartz, quartzite and lithic grains	4.4	5.1
Rhaetic	Mudstone, grey	1.4+	6.5

GRADING

Mean for Deposit

	%	mm	%
Gravel	8	+ 16	3
		- 16 + 4	5
Sand	72	- 4 + 1	5
		- 1 + $\frac{1}{4}$	38
		- $\frac{1}{4}$ + 1/16	29
Fines	20	- 1/16	20

Bulk Samples

Depth below surface (m)	Percentage Fines	Sand	Gravel
0.7 - 1.7	24	68	8
1.7 - 2.7	17	79	4
2.7 - 3.7	19	78	3
3.7 - 4.7	17	73	10
4.7 - 5.1	27	40	33

SK 89 NE 111 | 8606 9958 | Green Howe, Scotton | Block D

Surface level (+25.0 m) +82 ft
Water level (+23.9 m) +78 ft
February 1973

Overburden 0.3 m
Mineral 2.0 m
Waste 8.0 m
Bedrock 2.2 m+

LOG

Geological Classification	Lithology	Thickness m	Depth m
	Soil	0.3	0.3
Blown Sand	Sand: medium, subangular to rounded quartz and quartzite with chert and lithic grains	2.0	2.3
Boulder Clay	Silt, brown, with occasional subrounded pebbles of quartzite, chert, flint, chalk and mudstone	8.0	10.3
Lower Lias	Mudstone, pale brown	2.2+	12.5

GRADING

Mean for Deposit

	%	mm	%
Gravel	0	+ 16	0
		- 16 + 4	0
Sand	95	- 4 + 1	trace
		- 1 + $\frac{1}{4}$	61
		- $\frac{1}{4}$ + 1/16	34
Fines	5	- 1/16	5

Bulk Samples

Depth below surface (m)	Percentage Fines	Sand	Gravel
0.3 - 1.3	5	95	0
1.3 - 2.3	4	96	0

SK 89 NE 112	8624 9861	Laughton Wood, Scotton	Block D

Surface level (+22.6 m) +74 ft
Water not encountered
February 1973

Mineral 1.7 m
Waste 4.8 m
Bedrock 2.0 m+

LOG

Geological Classification	Lithology	Thickness m	Depth m
Blown Sand	'Very clayey' sand: fine, angular to subrounded quartz and quartzite with other lithic grains	1.7	1.7
Boulder Clay	Silt, pale brown, with occasional subrounded quartzite, chert, flint and chalk pebbles	4.8	6.5
Lower Lias	Mudstone, grey	2.0+	8.5

GRADING

Mean for Deposit

	%	mm	%
Gravel	trace	+ 16	0
		- 16 + 4	trace
Sand	72	- 4 + 1	1
		- 1 + $\frac{1}{4}$	26
		- $\frac{1}{4}$ + 1/16	45
Fines	28	- 1/16	28

Bulk Samples

Depth below surface (m)	Percentage Fines	Sand	Gravel
0.0 - 1.0	22	78	0
1.0 - 1.7	36	63	1

SK 89 NE 113 | 8694 9910 | Rainford Corner, Scotton | Block D

Surface level (+19.8 m) +65 ft
Water level (+17.8 m) +58 ft
February 1973

Overburden 0.2 m
Mineral 2.5 m
Waste 1.3 m
Bedrock 1.5 m+

LOG

Geological Classification	Lithology	Thickness m	Depth m
	Soil	0.2	0.2
Blown Sand	Sand: medium, subangular to rounded quartz and quartzite with chert and other lithic grains	2.5	2.7
Boulder Clay	Clay, pale brown mottled grey	1.3	4.0
Lower Lias	Mudstone, pale grey	1.5+	5.5

GRADING

Mean for Deposit

	%	mm	%
Gravel		+ 16	0
	trace	- 16 + 4	trace
		- 4 + 1	trace
Sand	95	- 1 + $\frac{1}{4}$	60
		- $\frac{1}{4}$ + 1/16	35
Fines	5	- 1/16	5

Bulk Samples

Depth below surface (m)	Percentage Fines	Sand	Gravel
0.2 - 1.2	4	96	0
1.2 - 2.2	7	93	0
2.2 - 2.7	1	99	trace

SK 89 NE 114 | 8733 9845 | Scotton | Block D

Surface level (+18.6 m) +61 ft
Water level (+16.6 m) +51 ft
February 1973

Overburden 0.3 m
Mineral 2.5 m
Bedrock 2.2 m+

LOG

Geological Classification	Lithology	Thickness m	Depth m
	Soil	0.3	0.3
Blown Sand	Sand, 'clayey' at base: medium, subangular to well rounded quartz, quartzite and other lithic grains	2.5	2.8
Lower Lias	Mudstone, grey	2.2+	5.0

GRADING

Mean for Deposit

	%	mm	%
Gravel	0	+16	0
		-16 + 4	0
Sand	94	- 4 + 1	trace
		- 1 + $\frac{1}{4}$	55
		- $\frac{1}{4}$ + 1/16	39
Fines	6	- 1/16	6

Bulk Samples

Depth below surface (m)	Percentage Fines	Sand	Gravel
0.3 - 1.3	4	96	0
1.3 - 2.3	5	95	0
2.3 - 2.8	12	88	0

SK 89 NE 115	8801 9957	Cockthorn Farm, Scotton	Block D

Surface level (+23.2 m) +76 ft
Water level (+20.0 m) +66 ft
January 1973

Overburden 0.5 m
Mineral 2.0 m
Bedrock 2.0 m+

LOG

Geological Classification	Lithology	Thickness m	Depth m
	Soil	0.5	0.5
Blown Sand, ? on Boulder Clay	'Very clayey' sand: fine, subangular to well rounded quartz, quartzite, chert and other lithic grains; lower sample probably contaminated with Boulder Clay	2.0	2.5
Lower Lias	Mudstone, grey	2.0+	4.5

GRADING

Mean for Deposit

	%	mm	%
Gravel	3	+ 16	2
		- 16 + 4	1
Sand	69	- 4 + 1	trace
		- 1 + $\frac{1}{4}$	26
		- $\frac{1}{4}$ + 1/16	43
Fines	28	- 1/16	28

Bulk Samples

Depth below surface (m)	Percentage Fines	Sand	Gravel
0.5 - 1.5	18	82	trace
1.5 - 2.5	38	57	5

SK 89 NE 116 8953 9961 Scotton Block D

Surface level (+8.2 m) +27 ft
Water level (+6.1 m) +20 ft
February 1973

Overburden 0.4 m
Mineral 3.0 m
Waste 1.1 m
Mineral 1.5 m
Bedrock 2.0 m+

LOG

Geological Classification	Lithology	Thickness m	Depth m
	Soil	0.4	0.4
Alluvium	(a) 'Very clayey' sand: medium, subangular to well rounded quartz and quartzite with some chert and other lithic grains; some chert gravel at base	3.0	3.4
	Clayey silt, grey	1.1	4.5
Terrace of River Eau	(b) Sandy gravel Gravel: medium, subrounded quartz and quartzite with subangular chert and flint and some rounded sandstone; abundant fossil shells Sand: medium, subangular to well rounded quartz, quartzite and lithic grains	1.5	6.0
Lower Lias	Mudstone, grey	2.0+	8.0

GRADING

Mean for Deposit

	%	mm	%
(a) Gravel	1	+ 16	0
		- 16 + 4	1
Sand	76	- 4 + 1	2
		- 1 + ¼	43
		- ¼ + 1/16	31
Fines	23	- 1/16	23
(b) Gravel	44	+ 16	8
		- 16 + 4	36
Sand	54	- 4 + 1	21
		- 1 + ¼	18
		- ¼ + 1/16	15
Fines	2	- 1/16	2

Bulk Samples

Depth below surface (m)	Percentage Fines	Sand	Gravel
0.4 - 1.4	40	60	trace
1.4 - 2.4	16	84	trace
2.4 - 3.4	15	83	2
4.5 - 5.5	1	52	47
5.5 - 6.0	2	60	38

SK 89 SW 27 | 8073 9469 | Cross Drain, East Stockwith | Block A

Surface level (+1.8 m) +6 ft
Water level not recorded
November 1972

Overburden 1.3 m
Mineral 5.2 m
Waste 0.3 m
Mineral 4.7 m+

LOG

Geological Classification		Lithology	Thickness m	Depth m
Alluvium		Peat and silt: 0.4 m brown silt on 0.9 m black peat	1.3	1.3
First Terrace	(a)	Sand: medium, angular to rounded quartz and quartzite with subangular chert	5.2	6.5
25-Ft Drift		Clay, brown, laminated	0.3	6.8
Older River Sand and Gravel	(b)	Sandy gravel on sand Gravel: fine, subangular to well rounded quartz and quartzite with chert, sandstone, some mudstone and siltstone and traces of limestone and coal Sand: medium, subangular to rounded quartz and rock fragments	4.7+	11.5
		Hole abandoned due to rising sand		

GRADING

Mean for Deposit

	%	mm	%
(a)			
Gravel	2	+ 16	0
		- 16 + 4	2
		- 4 + 1	1
Sand	96	- 1 + ¼	67
		- ¼ + 1/16	28
Fines	2	- 1/16	2
(b)			
Gravel	22	+ 16	8
		- 16 + 4	14
		- 4 + 1	4
Sand	77	- 1 + ¼	51
		- ¼ + 1/16	22
Fines	1	- 1/16	1

Bulk Samples

Depth below surface (m)	Percentage Fines	Sand	Gravel
1.3 - 2.3	6	94	trace
2.3 - 3.3	3	97	trace
3.3 - 4.3	1	93	6
4.3 - 5.3	1	99	trace
5.3 - 6.3	1	98	1
6.3 - 6.5	1	99	trace
6.8 - 7.8	1	43	56
7.8 - 8.8	1	89	10
8.8 - 9.8	1	68	31
9.8 - 10.8	1	95	4
10.8 - 11.5	1	98	1

COMPOSITION OF GRAVEL FRACTION
Percentages by weight (and number)

Depth below surface (m)	Quartzite	Quartz	Flint and chert	Sandstone	Mudstone and siltstone	Limestone	Ironstone	Coal
6.8 - 7.8	30(24)	29(30)	17(20)	17(20)	6(6)	trace	-	trace
7.8 - 8.8	29(22)	29(20)	22(25)	14(13)	2(6)	2(3)	trace	2(10)

SK 89 SW 28	8194 9466	Glebe Farm, Blyton	Block B

Surface level (+2.1 m) +7 ft
Water level (+0.6 m) +2 ft
December 1972

Overburden 1.8 m
Mineral 12.7 m
Bedrock 1.0 m+

LOG

Geological Classification	Lithology	Thickness m	Depth m
Alluvium	Soil on brown peaty silt	1.8	1.8
First Terrace on 25-Ft Drift and Older River Sand and Gravel	Sand, pebbly in lower part and 'clayey' from 4.8 to 6.8 m Gravel: fine, well rounded quartz and quartzite with sandstone and subangular chert Sand: fine to medium, well rounded quartz and rock fragments and some coal specks	12.7	14.5
Keuper Marl	Mudstone, red and green	1.0+	15.5

GRADING

Mean for Deposit

	%	mm	%
Gravel	2	+ 16	0
		- 16 + 4	2
Sand	94	- 4 + 1	2
		- 1 + $\frac{1}{4}$	54
		- $\frac{1}{4}$ + 1/16	38
Fines	4	- 1/16	4

Bulk Samples

Depth below surface (m)	Percentage Fines	Sand	Gravel
1.8 - 2.8	2	98	trace
2.8 - 3.8	1	98	1
3.8 - 4.8	7	93	trace
4.8 - 5.8	10	90	trace
5.8 - 6.8	15	85	trace
6.8 - 7.8	1	99	0
7.8 - 8.8	5	93	2
8.8 - 9.8	1	94	5
9.8 - 10.8	2	98	trace
10.8 - 11.8	1	98	1
11.8 - 12.8	1	97	2
12.8 - 13.8	1	84	15
13.8 - 14.5	3	91	6

SK 89 SW 29 | 8007 9368 | Cross Drain, East Stockwith | Block A

Surface level (+1.8 m) +6 ft
Water level (+0.8 m) +3 ft
November 1972

Overburden 1.0 m
Mineral 10.0 m
Bedrock 1.5 m+

LOG

Geological Classification	Lithology	Thickness m	Depth m
Alluvium	Soil on brown silt	1.0	1.0
	(a) Sand: medium, angular to rounded quartz, quartzite and other lithic grains; some fine rounded quartz and quartzite pebbles	6.0	7.0
Older River Sand and Gravel	(b) Sandy gravel Gravel: fine, subrounded to well rounded quartz, quartzite and sandstone, with angular to subrounded chert Sand: medium, angular to rounded quartz and rock fragments	4.0	11.0
Skerry in Keuper Marl	Sandstone, grey	1.5+	12.5

GRADING

Mean for Deposit

	%	mm	%
(a)			
Gravel	4	+ 16	2
		- 16 + 4	2
		- 4 + 1	2
Sand	87	- 1 + $\frac{1}{4}$	58
		- $\frac{1}{4}$ + 1/16	27
Fines	9	- 1/16	9
(b)			
Gravel	38	+ 16	8
		- 16 + 4	30
		- 4 + 1	13
Sand	61	- 1 + $\frac{1}{4}$	37
		- $\frac{1}{4}$ + 1/16	11
Fines	1	- 1/16	1

Bulk Samples

Depth below surface (m)	Percentage Fines	Sand	Gravel
1.0 - 2.0	8	87	5
2.0 - 3.0	20	77	3
3.0 - 4.0	10	90	trace
4.0 - 5.0	7	93	0
5.0 - 6.0	4	86	10
6.0 - 7.0	7	88	5
7.0 - 8.0	1	50	49
8.0 - 9.0	1	72	27
9.0 - 10.0	1	62	37
10.0 - 11.0	1	59	40

SK 89 SW 30	8025 9050	Gainsborough	Block A

Surface level (+3.0 m) +10 ft
Water level (+0.9 m) +3 ft
November 1972

Overburden 7.0 m
Mineral 1.0 m
Waste 2.0 m
Mineral 5.7 m
Bedrock 0.8 m+

LOG

Geological Classification	Lithology	Thickness m	Depth m
	Soil	0.2	0.2
Alluvium	Clay and peat: 2.8 m brown clay on 4.0 m brown clayey peat	6.8	7.0
Older River Sand and Gravel	(a) Gravel Gravel: fine to coarse, subrounded to well rounded quartz and quartzite with sandstone, and subangular chert Sand: medium, subangular to well rounded quartz and rock fragments	1.0	8.0
	No recovery	2.0	10.0
	(b) Sandy gravel Gravel: fine to coarse, subrounded quartz and quartzite, with subangular flint and chert Sand: medium, angular to rounded quartz, quartzite, chert and other lithic grains	5.7	15.7
Keuper Marl	Mudstone, green	0.8+	16.5

GRADING

Mean for Deposit

	%	mm	%
(a) Gravel	78	+ 16	39
		- 16 + 4	39
Sand	20	- 4 + 1	5
		- 1 + $\frac{1}{4}$	5
		- $\frac{1}{4}$ + 1/16	10
Fines	2	- 1/16	2
(b) Gravel	33	+ 16	5
		- 16 + 4	28
Sand	66	- 4 + 1	13
		- 1 + $\frac{1}{4}$	40
		- $\frac{1}{4}$ + 1/16	13
Fines	1	- 1/16	1

Bulk Samples

Depth below surface (m)	Percentages Fines	Sand	Gravel
7.0 - 8.0	2	20	78
10.0 - 11.0	1	94	5
11.0 - 12.0	trace	69	31
12.0 - 13.0	1	93	6
13.0 - 14.0	trace	45	55
14.0 - 15.0	1	43	56
15.0 - 15.7	1	37	62

SK 89 SW 31 | 8311 9496 | Owlett Plantation, Blyton | Block C

Surface level (+3.4 m) +11 ft
Water level (+0.4 m) +1 ft
December 1972

Overburden 0.4 m
Mineral 6.0 m
Waste 2.6 m
Mineral 3.8 m
Bedrock 2.2 m+

LOG

Geological Classification	Lithology	Thickness m	Depth m
	Soil	0.4	0.4
First Terrace	(a) Sand: fine to medium, subangular to well rounded quartz, quartzite and lithic grains	6.0	6.4
25-Ft Drift	Clay, brown, laminated	2.6	9.0
	(b) Sand, as above	3.8	12.8
Keuper Marl	Mudstone, grey, with gypsum in places	2.2+	15.0

GRADING

Mean for Deposit

	%	mm	%
(a)			
Gravel		+ 16	0
	trace	- 16 + 4	trace
		- 4 + 1	1
Sand	97	- 1 + $\frac{1}{4}$	52
		- $\frac{1}{4}$ + 1/16	44
Fines	3	- 1/16	3
(b)			
Gravel		+ 16	0
	trace	- 16 + 4	trace
		- 4 + 1	1
Sand	99	- 1 + $\frac{1}{4}$	45
		- $\frac{1}{4}$ + 1/16	53
Fines	1	- 1/16	1

Bulk Samples

Depth below surface (m)	Percentage Fines	Sand	Gravel
0.4 - 1.4	5	95	0
1.4 - 2.4	4	96	0
2.4 - 3.4	8	92	0
3.4 - 4.4	1	99	trace
4.4 - 5.4	1	99	trace
5.4 - 6.4	1	99	trace
9.0 - 10.0	1	99	trace
10.0 - 11.0	1	99	0
11.0 - 12.0	1	99	trace
12.0 - 12.8	3	97	trace

SK 89 SW 32 | 8404 9484 | Blyton | Block C

Surface level (+2.7 m) +9 ft
Water level not recorded
December 1972

Overburden 1.0 m
Mineral 2.0 m
Waste 2.9 m
Bedrock 1.1 m+

LOG

Geological Classification	Lithology	Thickness m	Depth m
Peat	Soil on dark brown and black peat	1.0	1.0
First Terrace	Sand: medium, angular to rounded quartz, quartzite and lithic grains	2.0	3.0
25-Ft Drift	Clay, brown	1.1	4.1
	Sand, as above, but fine-grained	0.4	4.5
	Clay, brown	1.4	5.9
Keuper Marl	Mudstone, grey	1.1+	7.0

GRADING

Mean for Deposit

	%	mm	%
Gravel	trace	+ 16	0
		- 16 + 4	trace
Sand	96	- 4 + 1	trace
		- 1 + $\frac{1}{4}$	63
		- $\frac{1}{4}$ + 1/16	33
Fines	4	- 1/16	4

Bulk Samples

Depth below surface (m)	Percentage Fines	Sand	Gravel
1.0 - 2.0	6	94	trace
2.0 - 3.0	1	98	1

SK 89 SW 33 | 8127 9400 | Stockwith Road, East Stockwith | Block B

Surface level (+3.4 m) +11 ft
Water level not recorded
November 1972

Overburden 0.4 m
Mineral 6.0 m
Waste 0.8 m
Mineral 7.8 m
Bedrock 1.5 m+

LOG

Geological Classification	Lithology	Thickness m	Depth m
	Soil	0.4	0.4
First Terrace	(a) Sand with 0.2 m clay band at 2.4 m: medium, subangular to rounded quartz, quartzite and other lithic grains	6.0	6.4
25-Ft Drift	Clay, brown and grey	0.8	7.2
25-Ft Drift on Older River Sand and Gravel	(b) Sand, pebbly below 11.5 m; 0.1 m clay band at 8.4 m Gravel: fine, subangular to well rounded quartz and quartzite, subangular chert and flint and rounded sandstone, with some mudstone and siltstone Sand: medium, subangular to rounded quartz, quartzite and other lithic grains	7.8	15.0
Keuper Marl	Mudstone, red and grey	1.5+	16.5

GRADING

Mean for Deposit

	%	mm	%
(a) Gravel	1	+ 16	0
		- 16 + 4	1
Sand	93	- 4 + 1	1
		- 1 + $\frac{1}{4}$	60
		- $\frac{1}{4}$ + 1/16	32
Fines	6	- 1/16	6
(b) Gravel	10	+ 16	2
		- 16 + 4	8
Sand	86	- 4 + 1	9
		- 1 + $\frac{1}{4}$	52
		- $\frac{1}{4}$ + 1/16	25
Fines	4	- 1/16	4

Bulk Samples

Depth below surface (m)	Percentage Fines	Sand	Gravel
0.4 - 2.4	3	97	trace
2.4 - 2.6	clay band*		
2.6 - 5.6	4	96	trace
5.6 - 6.4	1	96	3
7.2 - 8.2	9	89	2
8.2 - 8.4	2	98	trace
8.4 - 8.5	clay band*		
8.5 - 9.5	1	98	1
9.5 - 10.5	2	94	4
10.5 - 11.5	3	96	1
11.5 - 12.5	1	76	23
12.5 - 13.5	2	81	17
13.5 - 14.5	trace	77	23
14.5 - 15.0	3	81	16

* Assumed to comprise 100% fines in calculating mean grading.

COMPOSITION OF GRAVEL FRACTION
Percentages by weight (and number)

Depth below surface (m)	Quartzite	Quartz	Flint and chert	Sandstone	Mudstone and siltstone	Limestone
12.5 - 13.5	23(18)	23(27)	23(24)	20(22)	11(9)	trace
14.5 - 15.0	22(23)	39(26)	17(23)	12(13)	10(15)	-

SK 89 SW 34 8259 9418 Blyton Block B

Surface level (+3.0 m) +10 ft
Water level (-0.1 m) OD
December 1972

Overburden 0.5 m
Mineral 2.0 m
Waste 0.5 m
Mineral 2.5 m
Waste 1.0 m
Mineral 6.5 m
Bedrock 1.5 m+

LOG

Geological Classification	Lithology	Thickness m	Depth m
	Soil	0.5	0.5
First Terrace	(a) Sand: medium, subangular to rounded quartz, quartzite and lithic grains	2.0	2.5
	Silt, grey, slightly peaty	0.5	3.0
	(b) Sand: as above	2.5	5.5
25-Ft Drift	Clay, brown	1.0	6.5
	(c) Sand: as above but fine grained	6.5	13.0
Keuper Marl	Mudstone, red	1.5+	14.5

GRADING

Mean for Deposit

	%	mm	%
(a)			
Gravel	0	+ 16	0
		- 16 + 4	0
Sand	93	- 4 + 1	1
		- 1 + $\frac{1}{4}$	67
		- $\frac{1}{4}$ + 1/16	25
Fines	7	- 1/16	7
(b)			
Gravel	1	+ 16	0
		- 16 + 4	1
Sand	95	- 4 + 1	1
		- 1 + $\frac{1}{4}$	71
		- $\frac{1}{4}$ + 1/16	23
Fines	4	- 1/16	4
(c)			
Gravel	1	+ 16	trace
		- 16 + 4	1
Sand	95	- 4 + 1	2
		- 1 + $\frac{1}{4}$	41
		- $\frac{1}{4}$ + 1/16	52
Fines	4	- 1/16	4

Bulk Samples

Depth below surface (m)	Percentage Fines	Sand	Gravel
0.5 - 1.5	6	94	trace
1.5 - 2.5	8	92	0
3.0 - 4.0	2	96	2
4.0 - 5.0	4	96	trace
5.0 - 5.5	8	92	trace
6.5 - 7.5	3	95	2
7.5 - 8.5	12	88	trace
8.5 - 9.5	4	96	trace
9.5 - 10.5	3	95	2
10.5 - 11.5	1	98	1
11.5 - 12.5	1	97	2
12.5 - 13.0	1	96	3

SK 89 SW 35	8355 9395	Blyton	Block C

Surface level (+2.4 m) +8 ft
Water level (+1.4 m) +5 ft
December 1972

Overburden 1.5 m
Mineral 1.4 m
Waste 1.1 m
Mineral 4.1 m
Bedrock 1.4 m+

LOG

Geological Classification	Lithology	Thickness m	Depth m
Peat	Peat, dark brown and black	1.5	1.5
First Terrace	(a) Sand: medium, angular to rounded quartz, quartzite and other lithic grains	1.4	2.9
25-Ft Drift	Clay, brown	1.1	4.0
25-Ft Drift on Older River Sand and Gravel	(b) Sand, pebbly at base Gravel: mainly coarse, angular to subrounded chert Sand: medium, subrounded quartz, quartzite and other lithic grains	4.1	8.1
Keuper Marl	Mudstone, red	1.4+	9.5

GRADING

Mean for Deposit

	%	mm	%
(a)			
Gravel		+ 16	0
	trace	- 16 + 4	trace
		- 4 + 1	1
Sand	93	- 1 + ¼	60
		- ¼ + 1/16	32
Fines	7	- 1/16	7
(b)			
Gravel	5	+ 16	3
		- 16 + 4	2
		- 4 + 1	5
Sand	94	- 1 + ¼	55
		- ¼ + 1/16	34
Fines	1	- 1/16	1

Bulk Samples

Depth below surface (m)	Percentage Fines	Sand	Gravel
1.5 - 2.5	7	93	trace
2.5 - 2.9	4	96	trace
4.0 - 5.0	trace	98	2
5.0 - 6.0	trace	99	1
6.0 - 7.0	1	97	2
7.0 - 8.1	1	85	14

SK 89 SW 36 | 8070 9337 | Jarvis Hill, Morton | Block B

Surface level (+3.4 m) +11 ft
Water level (+1.9 m) +6 ft
November 1972

Overburden 0.4 m
Mineral 5.6 m
Waste 1.8 m
Mineral 7.0 m
Bedrock 1.7 m+

LOG

Geological Classification	Lithology	Thickness m	Depth m
	Soil	0.4	0.4
First Terrace	(a) Sand: fine to medium, subangular to well rounded quartz, quartzite and other lithic grains	5.6	6.0
25-Ft Drift	Clay, pale brown, with thin sand at 7.4 m	1.8	7.8
25-Ft Drift on Older River Sand and Gravel	(b) Sand and pebbly sand Gravel: fine to coarse, subangular to well rounded quartz and quartzite, with subangular chert and some rounded sandstone Sand: medium, subrounded quartz, quartzite with chert and other lithic grains	7.0	14.8
Keuper Marl	Mudstone, red	1.7+	16.5

GRADING

Mean for Deposit

	%	mm	%
(a)			
Gravel	1	+ 16	0
		- 16 + 4	1
		- 4 + 1	1
Sand	96	- 1 + ¼	52
		- ¼ + 1/16	43
Fines	3	- 1/16	3
(b)			
Gravel	9	+ 16	2
		- 16 + 4	7
		- 4 + 1	5
Sand	89	- 1 + ¼	48
		- ¼ + 1/16	36
Fines	2	- 1/16	2

Bulk Samples

Depth below surface (m)	Percentage Fines	Sand	Gravel
0.4 - 1.4	2	98	0
1.4 - 2.4	5	95	0
2.4 - 3.4	3	94	3
3.4 - 4.4	5	95	0
4.4 - 5.4	3	97	0
5.4 - 6.0	2	97	1
7.8 - 8.8	1	97	2
8.8 - 9.8	5	94	1
9.8 - 10.8	2	95	3
10.8 - 11.8	3	80	17
11.8 - 12.8	2	78	20
12.8 - 13.8	2	93	5
13.8 - 14.8	1	85	14

SK 89 SW 37 | 8206 9332 | Morton | Block B

Surface level (+3.0 m) +10 ft
Water level not recorded
November 1972

Overburden 0.4 m
Mineral 4.7 m
Waste 0.7 m
Mineral 5.9 m
Bedrock 1.3 m+

LOG

Geological Classification	Lithology	Thickness m	Depth m
	Soil	0.4	0.4
First Terrace	(a) Sand, 'very clayey' at 1.4 m: fine to medium, subangular to rounded quartz and quartzite with coal and other lithic grains	4.7	5.1
25-Ft Drift	Clay, grey and brown	0.7	5.8
	(b) Sand, 'clayey' at base: as above	5.9	11.7
Keuper Marl	Mudstone, red	1.3+	13.0

GRADING

Mean for Deposit

	%	mm	%
(a)			
Gravel		+ 16	0
	trace	- 16 + 4	trace
		- 4 + 1	trace
Sand	93	- 1 + $\frac{1}{4}$	54
		- $\frac{1}{4}$ + 1/16	39
Fines	7	- 1/16	7
(b)			
Gravel	1	+ 16	trace
		- 16 + 4	trace
		- 4 + 1	1
Sand	94	- 1 + $\frac{1}{4}$	38
		- $\frac{1}{4}$ + 1/16	55
Fines	5	- 1/16	5

Bulk Samples

Depth below surface (m)	Percentage Fines	Sand	Gravel
0.4 - 1.4	5	95	trace
1.4 - 2.4	24	76	0
2.4 - 3.4	1	99	trace
3.4 - 4.4	1	99	0
4.4 - 5.1	2	97	1
5.8 - 6.8	2	98	0
6.8 - 7.8	3	97	0
7.8 - 8.8	2	98	0
8.8 - 9.8	2	98	0
9.8 - 10.8	8	89	3
10.8 - 11.7	17	82	1

SK 89 SW 38   8311 9291   Thonock   Block C

Surface level (+5.5 m) +18 ft
Water level not recorded
December 1972

Mineral 2.3 m
Bedrock 2.7 m+

LOG

Geological Classification	Lithology	Thickness m	Depth m
First Terrace	Sand: medium, angular to rounded quartz, quartzite and other lithic grains	2.3	2.3
Keuper Marl	Mudstone, grey	2.7+	5.0

GRADING

Mean for Deposit

	%	mm	%
Gravel		+ 16	0
	trace	- 16 + 4	trace
		- 4 + 1	1
Sand	98	- 1 + ¼	77
		- ¼ + 1/16	20
Fines	2	- 1/16	2

Bulk Samples

Depth below surface (m)	Percentage Fines	Sand	Gravel
0.0 - 1.0	2	98	trace
1.0 - 2.0	1	99	0
2.0 - 2.3	1	99	trace

SK 89 SW 39 | 8418 9324 | Wharton | Block C

Surface level (+4.0 m) +13 ft
Water level (+1.8 m) +6 ft
December 1972

Overburden 0.5 m
Mineral 3.1 m
Bedrock 1.4 m+

LOG

Geological Classification	Lithology	Thickness m	Depth m
	Soil	0.2	0.2
First Terrace	Silt, pale brown, mottled black and yellow	0.3	0.5
	Sand: medium, angular to rounded quartz, quartzite and other lithic grains	3.1	3.6
Keuper Marl	Mudstone, red	1.4+	5.0

GRADING

Mean for Deposit

	%	mm	%
Gravel		+ 16	0
	trace	- 16 + 4	trace
		- 4 + 1	1
Sand	95	- 1 + $\frac{1}{4}$	63
		- $\frac{1}{4}$ + 1/16	31
Fines	5	- 1/16	5

Bulk Samples

Depth below surface (m)	Percentage Fines	Sand	Gravel
0.5 - 1.5	5	95	0
1.5 - 2.5	2	98	trace
2.5 - 3.6	7	93	trace

SK 89 SW 40 | 8032 9209 | Morton | Block C

Surface level (+3.4 m) +11 ft
Water level (+1.4 m) +6 ft
November 1972

Overburden 0.4 m
Mineral 4.0 m
Bedrock 1.6 m+

LOG

Geological Classification	Lithology	Thickness m	Depth m
	Soil	0.4	0.4
First Terrace	Sand: medium, angular to rounded quartz and quartzite with chert and other lithic grains	4.0	4.4
Keuper Marl	Mudstone, red	1.6+	6.0

GRADING

Mean for Deposit

	%	mm	%
Gravel		+ 16	0
	trace	- 16 + 4	trace
		- 4 + 1	2
Sand	95	- 1 + $\frac{1}{4}$	79
		- $\frac{1}{4}$ + 1/16	14
Fines	5	- 1/16	5

Bulk Samples

Depth below surface (m)	Percentage Fines	Sand	Gravel
0.4 - 1.4	7	93	trace
1.4 - 2.4	1	99	0
2.4 - 3.4	5	95	0
3.4 - 4.4	9	90	1

SK 89 SW 41	8124 9260	Morton	Block C

Surface level (+4.3 m) +14 ft
Water level (+2.5 m) +8 ft
November 1972

Overburden 0.3 m
Mineral 4.7 m
Waste 1.8 m
Mineral 1.0 m
Bedrock 2.2 m+

LOG

Geological Classification	Lithology	Thickness m	Depth m
	Soil	0.3	0.3
First Terrace	(a) Sand: medium, subangular to rounded quartz, quartzite, chert and other lithic grains	4.7	5.0
25-Ft Drift	Clay, pale brown	1.8	6.8
	(b) Sand: as above	1.0	7.8
Keuper Marl	Mudstone, red and grey	2.2+	10.0

GRADING

Mean for Deposit

	%	mm	%
(a)			
Gravel		+ 16	0
	trace	- 16 + 4	trace
		- 4 + 1	1
Sand	96	- 1 + ¼	69
		- ¼ + 1/16	26
Fines	4	- 1/16	4
(b)			
Gravel		+ 16	0
	trace	- 16 + 4	trace
		- 4 + 1	2
Sand	98	- 1 + ¼	48
		- ¼ + 1/16	48
Fines	2	- 1/16	2

Bulk Samples

Depth below surface (m)	Percentage Fines	Sand	Gravel
0.3 - 1.3	1	99	0
1.3 - 2.3	1	99	0
2.3 - 3.3	4	96	0
3.3 - 4.3	11	89	0
4.3 - 5.0	1	97	2
6.8 - 7.8	2	98	trace

SK 89 SW 42	8227 9252	Thonock	Block C

Surface level (+3.4 m) +11 ft
Water not encountered
December 1972

Waste 2.8 m
Bedrock 1.7 m+

LOG

Geological Classification	Lithology	Thickness m	Depth m
Peat	Peat, brown	1.4	1.4
First Terrace	Silt, pale brown	1.4	2.8
Keuper Marl	Mudstone, red	1.7+	4.5

SK 89 SW 43	8131 9248	Morton	Block C

Surface level (+3.4 m) +11 ft
Water not encountered
December 1972

Waste 1.0 m
Bedrock 3.0 m+

LOG

Geological Classification	Lithology	Thickness m	Depth m
	Soil	0.4	0.4
First Terrace	'Very clayey' sand: medium, angular to rounded quartz, quartzite, chert and rock fragments	0.6	1.0
Keuper Marl	Mudstone, grey	3.0+	4.0

GRADING

Mean for Deposit

	%	mm	%
Gravel	1	+ 16	0
		- 16 + 4	1
Sand	72	- 4 + 1	2
		- 1 + $\frac{1}{4}$	58
		- $\frac{1}{4}$ + 1/16	12
Fines	27	- 1/16	27

Bulk Samples

Depth below surface (m)	Percentage Fines	Sand	Gravel
0.4 - 1.0	27	72	1

SK 89 SW 44 — 8498 9434 — Blyton — Block E

Surface level (+25.3 m) +83 ft
Water not encountered
December 1972

Waste 5.0 m
Bedrock 1.0 m+

LOG

Geological Classification	Lithology	Thickness m	Depth m
	Soil	0.4	0.4
Glacial Sand and Gravel	Sand: medium, subangular to well rounded quartz and quartzite and some other lithic grains	0.9	1.3
Boulder Clay	Silt, brown and pale grey, with scattered sand pockets and occasional fine limestone and sandstone pebbles	3.7	5.0
Rhaetic	Mudstone, green	1.0+	6.0

GRADING

Mean for Deposit

	%	mm	%
Gravel	2	+ 16	0
		- 16 + 4	2
Sand	91	- 4 + 1	1
		- 1 + $\frac{1}{4}$	68
		- $\frac{1}{4}$ + 1/16	22
Fines	7	- 1/16	7

Bulk Samples

Depth below surface (m)	Percentage Fines	Sand	Gravel
0.4 - 1.3	7	91	2

SK 89 SW 45 | 8430 9385 | Blyton | Block E

Surface level (+5.8 m) +19 ft
Water level (+3.7 m) +12 ft
December 1972

Overburden 0.3 m
Mineral 2.0 m
Bedrock 1.7 m+

LOG

Geological Classification	Lithology	Thickness m	Depth m
	Soil	0.3	0.3
Glacial Sand and Gravel	Sand: medium, subangular to well rounded quartz, quartzite and other lithic grains	2.0	2.3
Keuper Marl	Mudstone, red	1.7+	4.0

GRADING

Mean for Deposit

	%	mm	%
Gravel		+ 16	0
	trace	- 16 + 4	trace
		- 4 + 1	trace
Sand	99	- 1 + $\frac{1}{4}$	78
		- $\frac{1}{4}$ + 1/16	21
Fines	1	- 1/16	1

Bulk Samples

Depth below surface (m)	Percentage Fines	Sand	Gravel
0.3 - 1.3	2	98	trace
1.3 - 2.3	1	99	trace

SK 89 SE 116 | 8711 9365 | Pilham | Block E

Surface level (+16.8 m) +55 ft
Water level (+13.8 m) +45 ft
December 1972

Overburden 2.5 m
Mineral 2.5 m
Waste 0.5 m
Mineral 1.0 m
Waste 1.5 m
Bedrock 1.0 m+

LOG

Geological Classification	Lithology	Thickness m	Depth m
	Soil	0.3	0.3
Boulder Clay	Silt, brown, with scattered pockets of sand and occasional quartzite, siltstone and flint pebbles	2.2	2.5
Glacial Sand and Gravel	(a) Sand: medium, subangular to rounded quartz, quartzite and other lithic grains	2.5	5.0
	Clay, pale brown	0.5	5.5
	(b) Sand: as above	1.0	6.5
Boulder Clay	Clay and silt, pale brown and grey with thin sandy gravel at 7.0 m	1.5	8.0
Lower Lias	Mudstone, dark grey	1.0+	9.0

GRADING

Mean for Deposit

	%	mm	%
(a)			
Gravel		+ 16	0
	trace	- 16 + 4	trace
		- 4 + 1	1
Sand	91	- 1 + $\frac{1}{4}$	61
		- $\frac{1}{4}$ + 1/16	29
Fines	9	- 1/16	9
(b)			
Gravel	0	+ 16	0
		- 16 + 4	0
		- 4 + 1	0
Sand	92	- 1 + $\frac{1}{4}$	72
		- $\frac{1}{4}$ + 1/16	20
Fines	8	- 1/16	8

Bulk Samples

Depth below surface (m)	Percentage Fines	Sand	Gravel
2.5 - 3.5	11	89	trace
3.5 - 4.5	10	90	0
4.5 - 5.0	1	98	1
5.5 - 6.5	8	92	0

SK 89 SE 117	8642 9268	Aisby	Block E

Surface level (+20.7 m) +68 ft
Water level (+17.9 m) +59 ft
December 1972

Overburden 4.9 m
Mineral 2.0 m
Waste 1.0 m
Bedrock 1.6 m+

LOG

Geological Classification	Lithology	Thickness m	Depth m
	Soil	0.2	0.2
Boulder Clay	Silt, brown, with abundant quartz, sandstone, flint and chalk pebbles	4.7	4.9
Glacial Sand and Gravel	'Clayey' sand: fine, angular to rounded quartz, quartzite and lithic grains	2.0	6.9
Boulder Clay	Silt, brown and grey; few quartz pebbles	1.0	7.9
Lower Lias	Mudstone, dark grey	1.6+	9.5

GRADING

Mean for Deposit

	%	mm	%
Gravel	0	+ 16	0
		- 16 + 4	0
Sand	90	- 4 + 1	trace
		- 1 + $\frac{1}{4}$	26
		- $\frac{1}{4}$ + 1/16	64
Fines	10	- 1/16	10

Bulk Samples

Depth below surface (m)	Percentage Fines	Sand	Gravel
4.9 - 5.9	9	91	0
5.9 - 6.9	12	88	trace

SK 89 SE 118 | 8744 9254 | Corringham | Block E

Surface level (+19.5 m) +64 ft
Water level (+16.4 m) +54 ft
December 1972

Overburden 0.3 m
Mineral 5.0 m
Waste 5.7 m
Bedrock 1.0 m+

LOG

Geological Classification	Lithology	Thickness m	Depth m
	Soil	0.3	0.3
Glacial Sand and Gravel	'Clayey' sand: fine to medium, subangular to rounded quartz, quartzite and other lithic grains	5.0	5.3
Boulder Clay	Silt, brown; thin 'clayey' pebbly sand bed at 9.3 m	5.7	11.0
Lower Lias	Mudstone, grey	1.0+	12.0

GRADING

Mean for Deposit

	%	mm	%
Gravel	2	+ 16	1
		- 16 + 4	1
Sand	86	- 4 + 1	1
		- 1 + $\frac{1}{4}$	39
		- $\frac{1}{4}$ + 1/16	46
Fines	12	- 1/16	12

Bulk Samples

Depth below surface (m)	Percentage Fines	Sand	Gravel
0.3 - 1.3	17	75	8
1.3 - 2.3	10	90	0
2.3 - 3.3	12	88	0
3.3 - 4.3	4	96	trace
4.3 - 5.3	17	83	trace

SK 89 SE 119 | 8640 9154 | Corringham | Block E

Surface level (+23.8 m) +78 ft
Water level (+20.6 m) +68 ft
December 1972

Overburden 3.6 m
Mineral 3.9 m
Waste 2.1 m
Bedrock 1.4 m+

LOG

Geological Classification	Lithology	Thickness m	Depth m
	Soil	0.2	0.2
Boulder Clay	Silt, brown, with scattered chalk, flint, siltstone and quartzite pebbles	3.4	3.6
Glacial Sand and Gravel	Sand: fine to medium, subangular to rounded quartz, quartzite and other lithic grains	3.9	7.5
Boulder Clay	Silt, brown and grey	2.1	9.6
Lower Lias	Mudstone, grey	1.4+	11.0

GRADING

Mean for Deposit

	%	mm	%
Gravel		+ 16	0
	trace	- 16 + 4	trace
		- 4 + 1	1
Sand	99	- 1 + $\frac{1}{4}$	38
		- $\frac{1}{4}$ + 1/16	60
Fines	1	- 1/16	1

Bulk Samples

Depth below surface (m)	Percentage Fines	Sand	Gravel
3.6 - 4.6	1	98	1
4.6 - 5.6	1	99	trace
5.6 - 6.6	1	99	0
6.6 - 7.5	1	99	0

SK 89 SE 120 | 8739 9104 | Corringham | Block E

Surface level (+17.4 m) +57 ft
Water level (+15.2 m) +50 ft
December 1972

Waste 4.0 m
Bedrock 1.5 m+

LOG

Geological Classification	Lithology	Thickness m	Depth m
	Soil	0.3	0.3
Boulder Clay	Silt and clay, brown, with abundant chalk, flint, siltstone, sandstone and quartzite pebbles; 0.3 m sand parting at 2.5 m	3.7	4.0
Lower Lias	Mudstone, grey	1.5+	5.5

APPENDIX G: CONVERSION TABLE, METRES TO FEET (TO NEAREST 0.5 FT)

m	ft	m	ft	m	ft	m	ft	m	ft
0.1	0.5	6.1	20	12.1	39.5	18.1	59.5	24.1	79
0.2	0.5	6.2	20.5	12.2	40	18.2	59.5	24.2	79.5
0.3	1	6.3	20.5	12.3	40.5	18.3	60	24.3	79.5
0.4	1.5	6.4	21	12.4	40.5	18.4	60.5	24.4	80
0.5	1.5	6.5	21.5	12.5	41	18.5	60.5	24.5	80.5
0.6	2	6.6	21.5	12.6	41.5	18.6	61	24.6	80.5
0.7	2.5	6.7	22	12.7	41.5	18.7	61.5	24.7	81
0.8	2.5	6.8	22.5	12.8	42	18.8	61.5	24.8	81.5
0.9	3	6.9	22.5	12.9	42.5	18.9	62	24.9	81.5
1.0	3.5	7.0	23	13.0	42.5	19.0	62.5	25.0	82
1.1	3.5	7.1	23.5	13.1	43	19.1	62.5	25.1	82.5
1.2	4	7.2	23.5	13.2	43.5	19.2	63	25.2	82.5
1.3	4.5	7.3	24	13.3	43.5	19.3	63.5	25.3	83
1.4	4.5	7.4	24.5	13.4	44	19.4	63.5	25.4	83.5
1.5	5	7.5	24.5	13.5	44.5	19.5	64	25.5	83.5
1.6	5	7.6	25	13.6	44.5	19.6	64.5	25.6	84
1.7	5.5	7.7	25.5	13.7	45	19.7	64.5	25.7	84.5
1.8	6	7.8	25.5	13.8	45.5	19.8	65	25.8	84.5
1.9	6	7.9	26	13.9	45.5	19.9	65.5	25.9	85
2.0	6.5	8.0	26	14.0	46	20.0	65.5	26.0	85.5
2.1	7	8.1	26.5	14.1	46.5	20.1	66	26.1	85.5
2.2	7	8.2	27	14.2	46.5	20.2	66.5	26.2	86
2.3	7.5	8.3	27	14.3	47	20.3	66.5	26.3	86.5
2.4	8	8.4	27.5	14.4	47	20.4	67	26.4	86.5
2.5	8	8.5	28	14.5	47.5	20.5	67.5	26.5	87
2.6	8.5	8.6	28	14.6	48	20.6	67.5	26.6	87.5
2.7	9	8.7	28.5	14.7	48	20.7	68	26.7	87.5
2.8	9	8.8	29	14.8	48.5	20.8	68	26.8	88
2.9	9.5	8.9	29	14.9	49	20.9	68.5	26.9	88.5
3.0	10	9.0	29.5	15.0	49	21.0	69	27.0	88.5
3.1	10	9.1	30	15.1	49.5	21.1	69	27.1	89
3.2	10.5	9.2	30	15.2	50	21.2	69.5	27.2	89
3.3	11	9.3	30.5	15.3	50	21.3	70	27.3	89.5
3.4	11	9.4	31	15.4	50.5	21.4	70	27.4	90
3.5	11.5	9.5	31	15.5	51	21.5	70.5	27.5	90
3.6	12	9.6	31.5	15.6	51	21.6	71	27.6	90.5
3.7	12	9.7	32	15.7	51.5	21.7	71	27.7	91
3.8	12.5	9.8	32	15.8	52	21.8	71.5	27.8	91
3.9	13	9.9	32.5	15.9	52	21.9	72	27.9	91.5
4.0	13	10.0	33	16.0	52.5	22.0	72	28.0	92
4.1	13.5	10.1	33	16.1	53	22.1	72.5	28.1	92
4.2	14	10.2	33.5	16.2	53	22.2	73	28.2	92.5
4.3	14	10.3	34	16.3	53.5	22.3	73	28.3	93
4.4	14.5	10.4	34	16.4	54	22.4	73.5	28.4	93
4.5	15	10.5	34.5	16.5	54	22.5	74	28.5	93.5
4.6	15	10.6	35	16.6	54.5	22.6	74	28.6	94
4.7	15.5	10.7	35	16.7	55	22.7	74.5	28.7	94
4.8	15.5	10.8	35.5	16.8	55	22.8	75	28.8	94.5
4.9	16	10.9	36	16.9	55.5	22.9	75	28.9	95
5.0	16.5	11.0	36	17.0	56	23.0	75.5	29.0	95
5.1	17	11.1	36.5	17.1	56	23.1	76	29.1	95.5
5.2	17	11.2	36.5	17.2	56.5	23.2	76	29.2	96
5.3	17.5	11.3	37	17.3	57	23.3	76.5	29.3	96
5.4	17.5	11.4	37.5	17.4	57	23.4	77	29.4	96.5
5.5	18	11.5	37.5	17.5	57.5	23.5	77	29.5	97
5.6	18.5	11.6	38	17.6	57.5	23.6	77.5	29.6	97
5.7	18.5	11.7	38.5	17.7	58	23.7	78	29.7	97.5
5.8	19	11.8	38.5	17.8	58.5	23.8	78	29.8	98
5.9	19.5	11.9	39	17.9	58.5	23.9	78.5	29.9	98
6.0	19.5	12.0	39.5	18.0	59	24.0	78.5	30.0	98.5

REFERENCES

ALLEN, V.T. 1936. Terminology of medium-grained sediments. Rep. Natl. Res. Counc. Washington, 1935-1936, App. 1. Rep. Comm. Sedimentation, pp. 18-47.

ARCHER, A.A. 1969. Background and problems of an assessment of sand and gravel resources in the United Kingdom. Pp.495-508 in Proc. 9th Commonw. Min. Metall. Congr. 1969, Vol. 2, Mining and Petroleum Geology. (London: The Institute of Mining and Metallurgy).

——— 1970a. Standardisation of the size classification of naturally occurring particles. Géotechnique, Vol. 20, pp. 103-207.

——— 1970b. Making the most of metrication. Quarry Manager's J., Vol. 54, pp. 223-227.

ATTERBERG, A. 1905. Die rationelle Klassifikation der Sande und Kiese. Chem. Z., Vol. 29, pp. 195-198.

BRITISH STANDARD 1377. 1967. Methods of Testing Soils for civil engineering purposes. (London: British Standards Institution). 233 pp.

BUREAU OF MINES AND GEOLOGICAL SURVEY. 1948. Mineral Resources of the United States. (Washington, D.C.: Public Affairs Press), pp. 14-17.

EDWARDS, W. 1936. A Pleistocene strand-line in the Vale of York. Proc. Yorkshire Geol. Soc., Vol. 23, pp. 103-118.

GAUNT, G.D. 1974. A radiocarbon date relating to Lake Humber. Proc. Yorkshire Geol. Soc., Vol. 40, pp. 195-197.

——— JARVIS, R.A. and MATTHEWS, S.B. 1971. The late Weichselian sequence in the Vale of York. Proc. Yorkshire Geol. Soc., Vol. 38, pp. 281-284.

HARRIS, P.M., THURRELL, R.G., HEALING, R.A. and ARCHER, A.A. 1974. Aggregates in Britain. Proc. R. Soc. London, Ser. A, Vol. 399, pp. 329-353.

LANE, E.W. and others. 1947. Report of the sub-committee on sediment terminology. Trans. Am. Geophys. Un., Vol. 28, pp. 936-938.

PETTIJOHN, F.G. 1957. Sedimentary Rocks. 2nd Ed. (London: Harper and Row).

SMITH, E.G., RHYS, G.H. and GOOSENS, R.F. 1973. Geology of the country around East Retford, Worksop and Gainsborough. Mem. Geol. Surv. G.B.

THURRELL, R.G. 1971. The assessment of mineral resources with particular reference to sand and gravel. Quarry Manager's J., Vol. 55, pp. 19-25.

TWENHOFEL, W.H. 1937. Terminology of the fine-grained mechanical sediments. Rep. Natl. Counc.Washington 1936-1937, App. 1, Rep. Comm. Sedimentation, pp. 81-104.

UDDEN. J.A. 1914. Mechanical composition of clastic sediments. Bull. Geol. Soc. Am., Vol. 25, pp. 655-744.

WENTWORTH, C.K. 1922. A scale of grade and class terms for clastic sediments. J. Geol., Vol. 30, pp. 377-392.

——— 1935. The terminology of coarse sediments. Bull. Natl. Res. Counc. Washington, No. 98, pp. 225-246.

WILLMAN, H.B. 1942. Geology and mineral resources of the Marseilles, Ottawa and Streatar quadrangles. Bull. Illinois State Geol. Surv., No. 66, pp. 343-344.

Dd. 587334 K 8
Printed in England for Her Majesty's Stationery Office
by Commercial Colour Press, London E7

The following reports of the Institute relate particularly to bulk mineral resources

Reports of the Institute of Geological Sciences

Assessment of British Sand and Gravel Resources

1 The sand and gravel resources of the country south-east of Norwich, Norfolk: Resource sheet TG 20. E. F. P. Nickless. Report 71/20 ISBN 0 11 880216 £1.15

2 The sand and gravel resources of the country around Witham, Essex: Resource sheet TL 81. H. J. E. Haggard. Report 72/6 ISBN 0 11 880588 6 £1.20

3 The sand and gravel resources of the area south and west of Woodbridge, Suffolk: Resource sheet TM 24. R. Allender and S. E. Hollyer.
Report 72/9 ISBN 0 11 880596 7 £1.70

4 The sand and gravel resources of the country around Maldon, Essex: Resource sheet TL 80. J. D. Ambrose. Report 73/1 ISBN 0 11 880600 9 £1.20

5 The sand and gravel resources of the country around Hethersett, Norfolk: Resource sheet TG 10.
E. F. P. Nickless.
Report 73/4 ISBN 0 11 880606 8 £1.60

6 The sand and gravel resources of the country around Terling, Essex: Resource sheet TL 71. C. H. Eaton. Report 73/5 ISBN 0 11 880608 4 £1.20

7 The sand and gravel resources of the country around Layer Breton and Tolleshunt D'Arcy, Essex: Resource sheet TL 91 and part of TL 90. J. D. Ambrose.
Report 73/8 ISBN 0 11 990614 9 £1.30

8 The sand and gravel resources of the country around Shotley and Felixstowe, Suffolk: Resource sheet TM 23. R. Allender and S. E. Hollyer.
Report 73/13 ISBN 0 11 880625 4 £1.60

9 The sand and gravel resources of the country around Attlebridge, Norfolk: Resource sheet TG 11.
E. F. P. Nickless.
Report 73/15 ISBN 0 11 880658 0 £1.85

10 The sand and gravel resources of the country west of Colchester, Essex: Resource sheet TL 92. J. D. Ambrose. Report 74/6 ISBN 0 11 880671 8 £1.45

11 The sand and gravel resources of the country around Tattingstone, Suffolk: Resource sheet TM 13. S. E. Hollyer. Report 74/9 ISBN 0 11 880675 0 £1.95

12 The sand and gravel resources of the country around Gerrards Cross, Buckinghamshire: Resource sheets SU 99, TQ 08 and TQ 09. H. C. Squirrell.
Report 74/14 ISBN 0 11 880710 2 £2.20

Mineral Assessment Reports

13 The sand and gravel resources of the country east of Chelmsford, Essex: Resource sheet TL 70. M. R. Clark. ISBN 0 11 880744 7 £3.50

14 The sand and gravel resources of the country east of Colchester, Essex: Resource sheet TM 02. J. D. Ambrose. ISBN 0 11 880745 5 £3.25

15 The sand and gravel resources of the country around Newton on Trent, Lincolnshire: Resource sheet SK 87.
D. Price.
ISBN 0 11 880746 3 £3.00

16 The sand and gravel resources of the country around Braintree, Essex: Resource sheet TL 72. M. R. Clark. ISBN 0 11 880747 1 £3.50

17 The sand and gravel resources of the country around Besthorpe, Nottinghamshire: Resource sheet SK 86 and part of SK 76. J. R. Gozzard.
ISBN 0 11 880748 X £3.00

18 The sand and gravel resources of the Thames Valley, the country around Cricklade, Wiltshire: Resource sheets SU 09/19 and parts of SP 00/10. P. R. Robson.
ISBN 0 11 880749 8 £3.00

19 The sand and gravel resources of the country south of Gainsborough, Lincolnshire: Resource sheet SK 88 and part of SK 78. J. H. Lovell.
ISBN 0 11 880750 1 £2.50

20 The sand and gravel resources of the country east of Newark upon Trent, Nottinghamshire: Resource sheet SK 85 ISBN 0 11 880751 X £2.75

21 The sand and gravel resources of the Thames and Kennet Valleys, the country around Pangbourne, Berkshire: Resource sheet SU 67. H. C. Squirrell.
ISBN 0 11 880752 8 £3.25

22 The sand and gravel resources of the country north-west of Scunthorpe, Humberside: Resource sheet SE 81.
J. W. C. James.
ISBN 0 11 880753 6 £3.00

23 The sand and gravel resources of the Thames Valley, the country between Lechlade and Standlake: Resource sheet SP 30 and parts of SP 20, SU 29 and SU 39. P. Robson. ISBN 0 11 881252 1 £7.25

24 The sand and gravel resources of the country around Aldermaston, Berkshire: Parts of resource sheets SU 56 and SU 66. H. C. Squirrell.
ISBN 0 11 881253 X £5.00

25 The celestite resources of the area north-east of Bristol: Resource sheet ST 68 and parts of ST 59, 69, 79, 58, 78, 68 and 77. E. F. P. Nickless, S. J. Booth and P. N. Mosley.
ISBN 0 11 881262 9 £5.00

26 The limestone and dolomite resources of the country around Monyash, Derbyshire: Resource sheet SK 16.
F. C. Cox and D. McC. Bridge.
ISBN 0 11 881263 7 £7.00

27 The sand and gravel resources of the country west and south of Lincoln, Lincolnshire: Resource sheets SK 95, SK 96 and SK 97. I. Jackson.
ISBN 0 11 884003 7 £6.00

28 The sand and gravel resources of the country around Eynsham, Oxfordshire: Resource sheet SP 40 and part of SP 41. W. J. R. Harries.
ISBN 0 11 884012 6 £3.00

29 The sand and gravel resources of the country south-west of Scunthorpe, Humberside: Resource sheet SE 80.
J. H. Lovell.
ISBN 0 11 884013 4 £3.50

30 Procedure for the assessment of limestone resources. F. C. Cox, D. McC. Bridge and J. H. Hull.
ISBN 0 11 884030 4 £1.25

31 The sand and gravel resources of the country west of Newark upon Trent, Nottinghamshire. Resource sheet SK 75. D. Price and P. J. Rogers.
ISBN 0 11 884031 2 £3.50

32 The sand and gravel resources of the country around Sonning and Henley. Resource sheets SU 77 and SU 78.
H. C. Squirrell.
ISBN 0 11 884032 0 £5.25

33 The sand and gravel resources of the country north of Gainsborough. Resource sheet SK 89. J. Gozzard and D. Price.
ISBN 0 11 884033 9 £4.50

34 The sand and gravel resources of the Dengie Peninsula, Essex: Resource sheet TL 90, etc. M. B. Simmons.
ISBN 0 11 884081 9 *not yet priced*

35 The sand and gravel resources of the country around Darvel: Sheets NS 53, 63, etc. E. F. P. Nickless, A. M. Aitken and A. A. McMillan.
ISBN 0 11 884082 7 £7.00

36 The sand and gravel resources of the country around Southend-on-Sea, Essex: Resource sheets TQ 78/79 etc.
S. E. Hollyer and M. B. Simmons.
ISBN 0 11 884083 5 £7.50

37 The sand and gravel resources of the country around Bawtry, South Yorkshire: Resource sheet SK 69.
A. R. Clayton.
ISBN 0 11 884053 3 *not yet priced*

38 The sand and gravel resources of the country around Abingdon, Oxfordshire: Resource sheets SU 49, 59 and SP 40, 50. C. E. Corser.
ISBN 0 11 884084 3 £5.50

Reports of the Institute of Geological Sciences

Other Reports

69/9 Sand and gravel resources of the inner Moray Firth. A. L. Harris and J. D. Peacock.
ISBN 0 11 880106 6 35p

70/4 Sands and gravels of the southern counties of Scotland. G. A. Goodlet.
ISBN 0 11 880105 8 90p

72/8 The use and resources of moulding sand in Northern Ireland. R. A. Old.
ISBN 0 11 881594 0 30p

73/9 The superficial deposits of the Firth of Clyde and its sea lochs. C. E. Deegan, R. Kirby, I. Rae and R. Floyd.
ISBN 0 11 880617 3 95p

77/1 Sources of aggregate in Northern Ireland (2nd edition). I. B. Cameron.
ISBN 0 11 881279 3 70p

77/2 Sand and gravel resources of the Grampian Region. J. D. Peacock and others.
ISBN 0 11 881282 3 80p

77/5 Sand and gravel resources of the Fife Region. M. A. E. Browne.
ISBN 0 11 884004 5 60p

77/6 Sand and gravel resources of the Tayside Region. I. B. Paterson.
ISBN 0 11 884008 8 £1.40

77/8 Sand and gravel resources of the Strathclyde Region. I. B. Cameron and others.
ISBN 0 11 884028 2 £2.50

77/9 Sand and gravel resources of the Central Region, Scotland. M. A. E. Browne.
ISBN 0 11 884016 9 £1.35

77/19 Sand and gravel resources of the Borders Region of Scotland. A. D. McAdam.
ISBN 0 11 884025 8 £1.00

77/22 Sand and gravel resources of the Dumfries and Galloway Region of Scotland. I. B. Cameron.
ISBN 0 11 884021 5 £1.20

78/1 Sand and gravel resources of the Lothian Region of Scotland. A. D. McAdam.
ISBN 0 11 884042 8 £1.00